Cambridge Tracts in Mathematics
and Mathematical Physics

GENERAL EDITORS
J. F. C. KINGMAN, F. SMITHIES,
J. A. TODD, C. T. C. WALL,
AND H. BASS

No. 59

PROXIMITY SPACES

PROXIMITY
SPACES

S. A. NAIMPALLY
Professor of Mathematics
Indian Institute of Technology
Kanpur

AND

B. D. WARRACK
Department of Mathematics
University of Alberta

CAMBRIDGE
AT THE UNIVERSITY PRESS
1970

CAMBRIDGE UNIVERSITY PRESS
Cambridge, New York, Melbourne, Madrid, Cape Town, Singapore, São Paulo, Delhi

Cambridge University Press
The Edinburgh Building, Cambridge CB2 8RU, UK

Published in the United States of America by Cambridge University Press, New York

www.cambridge.org
Information on this title: www.cambridge.org/9780521079358

First published 1970
This digitally printed version 2008

A catalogue record for this publication is available from the British Library

Library of Congress Catalogue Card Number: 73–118858

ISBN 978-0-521-07935-8 hardback
ISBN 978-0-521-09183-1 paperback

CONTENTS

PREFACE

This tract aims at providing a compact introduction to the theory of proximity spaces and their generalizations. It is hoped that a study of the tract will better enable the reader to understand the current literature. In view of the fact that research material on proximity spaces is scattered and growing rapidly, the need for such a survey is apparent. The material herein is self-contained except for a basic knowledge of topological and uniform spaces, as can be found in standard texts such as the one by John L. Kelley; in fact, for the most part, we use Kelley's notation and terminology.

The tract begins with a brief history of the subject. The first two chapters give the fundamentals and the pace of development is rather slow. We have tried to motivate definitions and theorems with the help of metric and uniform spaces; a knowledge of the latter is, however, not necessary in understanding the proofs. The main result in these two chapters is the existence of the Smirnov compactification, which is proved using clusters. Taking advantage of hindsight, several proofs have been considerably simplified.

A reader not acquainted with uniform spaces will find it necessary to become familiar with such spaces before reading the third chapter. In this chapter, the interrelationships between proximity structures and uniform structures are considered and, since proximity spaces are intermediate between topological and uniform spaces, some of the most exciting results are to be found in this part of the tract. Various generalizations of uniform spaces find their way naturally into the theory presented here. The final chapter deals with several generalized forms of proximity structures, with one of them being studied in some detail. This chapter is rather sketchy and the interested reader is referred to the relevant literature for further information.

In order to minimize the number of discontinuities occurring in the main body of the text, all references from which material is selected as well as those where further details can be found are collected together in the Notes at the end of each chapter. We have attempted to provide a reasonably complete bibliography of the literature on proximity spaces; to this end we were greatly aided by D. Bushaw's *Bibliography on Uniform Topology* (Washington State University, November, 1965). At the end of each item in the bibliography is found a reference to Mathematical Reviews. Appended separately is a list of general references used in the tract. An index of notations and another of terms are also included.

With great pleasure we acknowledge our indebtedness to several colleagues. Dr K. M. Garg, Mr C. M. Pareek, Professor A. J. Ward and Professor K. Iséki assisted with advice during the initial stages. Comments by Professor C. T. C. Wall on the first draft of the manuscript were useful during revision. Several mathematicians kindly sent us their unpublished manuscripts; we are especially grateful to Dr C. J. Mozzochi, who also made several suggestions. Mathematical manuscripts are difficult to type and we admire the skill and patience of our typists: Miss June Talpash, Mrs Vivian Spak and Mrs Georgina Smith.

The first author would like to take this opportunity to express gratitude to his inspiring teachers: Professors D. S. Agashé, M. L. Chandratreya, D. P. Patravali, N. H. Phadke from India, and Professors J. G. Hocking and D. E. Sanderson from the U.S.A. This author was generously supported by operating grants from the National Research Council (Canada) and the Summer Research Institute of the Canadian Mathematical Congress (1967).

We thank the staff of the Cambridge University Press for their help and cooperation. Miss M. Gagrat helped in the difficult task of proof correction.

S. A. N.
B. D. W.

May 1969
The University of Alberta
Edmonton, Canada

INDEX OF NOTATIONS

[ix]

HISTORICAL BACKGROUND

The germ of the theory of proximity spaces showed itself as early as 1908 at the mathematical congress in Bologna, when Riesz [95] discussed various ideas in his 'theory of enchainment' which have today become the basic concepts of the theory. The subject was essentially rediscovered in the early 1950's by Efremovič [18, 19] when he axiomatically characterized the proximity relation 'A is near B' for subsets A and B of any set X. The set X together with this relation was called an infinitesimal (proximity) space, and is a natural generalization of a metric space and of a topological group. A decade earlier a study was made by Krishna Murti [52], Wallace [116, 117] and Szymanski [113] concerning the use of 'separation of sets' as the primitive concept. In each case similar, but weaker, axioms than those of Efremovič were used. Efremovič later used proximity neighbourhoods to obtain an equivalent set of axioms for a proximity space and thereby an alternative approach to the theory.

Defining the closure of a subset A of X to be the collection of all points of X 'near' A, Efremovič [19] showed that a topology can be introduced in a proximity space and that one thereby obtains, in fact, a completely regular (and hence uniformizable) space. He further showed that every completely regular space X can be turned into a proximity space with the help of Urysohn's function: namely, $A \, \delta \, B$ iff there exists a continuous function f mapping X into $[0, 1]$ such that $f(A) = 0$ and $f(B) = 1$. Smirnov [98] subsequently proved that every completely regular space has a maximal associated proximity space, and that it has a minimal associated proximity space if and only if it is locally compact.

More recently, Mrówka [75] has introduced a nearness relation on the set of all sequences from a proximity space. This provides a motivation for defining a notion, similar to that of proximity, in a Fréchet L-space; such a notion was discovered independently by Goetz [28] to obtain what he terms a $\mathscr{UL}$-space. However, as opposed to the complete regularity of proximity space, a $\mathscr{UL}$-space need not even be a topological space. Poljakov [91] and Goetz [29] have since carried out further investigations in this area, and discuss the connection between this notion and the proximity of Efremovič. Švarc [20] had earlier introduced a nearness relation on the set of all nets from a given space.

In order to study mappings from one proximity space to another, it was natural for Efremovič to introduce the concept of a proximity mapping. Defined to be a mapping which preserves the proximity of sets, a proximity mapping is a natural analogue of a continuous mapping in topological spaces and of a uniformly continuous mapping in uniform spaces. It is readily verified that a proximity mapping between two proximity spaces is continuous with respect to the induced topologies. Pervin [86] later revealed that the converse holds if the domain proximity space is equinormal.

In 1952, Smirnov [98] pursued extensions of proximity spaces and in particular answered Alexandroff's query: 'which topological spaces admit a proximity relation compatible with the given topology?'. He discovered the connection between the Hausdorff compactification of a Tychonoff space and the compatible proximity relation, showing that 'a topological space admits a compatible proximity relation if and only if it is a subspace of a compact Hausdorff space'. Using the 'ends' of Alexandroff [A], Smirnov [98] obtained the compactification of a proximity space X by identifying each point $x \in X$ with the end consisting of all proximity neighbourhoods of x, and showing the compactification of X to be the set of all ends in X.

The concept of a cluster, the analogue in a proximity space of an ultrafilter, was introduced by Leader [53] and provides an alternative approach to many proximity problems. In particular, Leader [53] obtained the compactification of a proximity space X to be the family of all clusters from X. Since the Smirnov compactification is unique, it is evident that a one-to-one correspondence exists between clusters and ends. In fact, Leader [55] proves (without resorting to compactification theory) that clusters and ends are dual classes.

Császár and Mrówka [13] proved a stronger result regarding the compactification of a proximity space: namely, compactification can be effected preserving the proximity weight. This gave rise to the following interesting metrization theorem: a proximity space of proximity weight $\aleph_0$ is metrizable. Additional solutions to the metrization problem were offered earlier by Efremovič and Švarc [20], who used the 'sequence-uniformity' method, and by Ramm and Švarc [92] using uniform proximity covers. Smirnov [105] has used both the pseudo-metric and proximity neighbourhood approaches in deriving necessary and sufficient conditions for metrizability. More recently, Leader [61] has investigated this problem in a manner analogous to R. L. Moore's approach to the metrization of topological spaces.

Using the alternate set of axioms for uniform spaces involving uniform coverings, Smirnov [98] showed that every proximity (or p-) equivalence class of uniform structures contains a coarsest member, which is also the unique totally bounded structure of the class. He also showed that there is an isomorphism between the partially ordered set of all proximities on a given completely regular space and the partially ordered set of all its compactifications, which reduces the theory of proximity spaces to that of compactifications.

In 1959, Gál [27] continued the pursuit of the relationships between uniform structures and proximities, proving that there is a natural order-preserving one-to-one correspondence between totally bounded (precompact) structures and proximity relations. Gál also showed that there is a one-to-one correspondence between the compactifications of a uniformizable space and the totally bounded uniform structures which are compatible with its topology. Again, this yields a one-to-one map between the Hausdorff compactifications and the separated structures. In the same year, Alfsen and Fenstad [4] paralleled the work of Gál and treated many of the problems of Efremovič and Smirnov in the framework of Weil's uniform structures. Using maximal regular filters Alfsen and Fenstad perform completion and compactification, showing that there exists a maximal completion amongst all the completions determined by the structures of a given proximity equivalence class. This naturally raises the question as to whether there always exists a minimal completion, which is equivalent to asking if there exists a finest uniform structure in a given proximity equivalence class. This question was answered affirmatively by Smirnov [98] for the case in which the proximity space is metrizable. That the answer is in general negative was first shown in 1961 by Dowker [17]. He unveiled an example of a proximity space (a product of two infinite spaces with the product proximity structure) which has no finest uniform structure inducing its proximity.

The key concept in any completion theory for proximity spaces is that of 'small' sets, which can be introduced by means of pseudo-metrics, uniform structures or uniform coverings. Leader [54] has used the first device to define a local cluster, while the authors [121] have used the second to define a Cauchy cluster. They obtain a completion which consists of the family of all local (resp. Cauchy) clusters in X, a subspace of the family of all clusters, which Leader showed to be the Smirnov compactification. In [99, 100], Smirnov used uniform δ-coverings to introduce the concept of a complete uniform space and proved that every proximity space admits a minimal completion of this kind. Thus the general existence of a minimal completion was established, but with the sacrifice of the one-to-one correspondence between proximity structures and completions.

This was regained in the work of Alfsen and Njåstad [6], who in 1963 gave yet another example of a p-equivalence class lacking a finest uniform structure. (Further examples are to be found in Leader [56], Fenstad [23] and Isbell [44].) This then led to the notion of a generalized uniform structure, obtained from Weil's uniform structure by replacing the 'intersection' axiom with a less restrictive one. They proved that for generalized uniform structures the answer to Smirnov's problem is affirmative, and gave an explicit characterization of those generalized uniform structures which occur as the finest member of their respective p-equivalence classes. Such members are called *total* structures, and it was shown that the collection of all total generalized uniform structures embraces all ordinary metrizable (or pseudometrizable) uniform structures. Alfsen and Njåstad also found a relation between proximal continuity and uniform continuity, obtaining as a particular consequence, Efremovič's result that metric uniform continuity is equivalent to metric proximal continuity. They further proved that every generalized uniform space can be completed, and established a one-to-one corre-spondence between generalized uniform structures and proximity space completions. In particular, the minimal completion of a proximity space is obtained by completion of the finest general-ized structure compatible with the proximity structure.

In the same year, Njåstad [80] further developed the concept of a generalized uniform structure. He first noted that the col-lection of all generalized uniform structures on a set and the collection of all proximity structures on a set form complete lattices when ordered by the relations 'finer-coarser', and that lattice sums and products are compatible, unlike the case in which the usual uniform structures are considered. Moreover, he established the existence and compatibility of initial (final) generalized uniform structures and initial (final) proximity structures. He proved that if the uniformity is replaced by the associated proximity, then uniform convergence implies con-vergence in proximity, a notion introduced by Leader [54]. Convergence in proximity implies uniform convergence if the associated total uniformity is used in place of the proximity. Moreover, for a net of functions with a linearly ordered directed set, the two forms of convergence are equivalent. In 1965, Hursch [37] formulated a new concept, *height*, to help clarify the order structure of p-equivalence classes of uniformities.

As previously mentioned, some authors have worked with weaker axioms than those of Efremovič, enabling them to intro-duce an arbitrary topology on the underlying set. With such generalized proximities as quasi-proximity, paraproximity, pseudo-proximity and local proximity already existing in the literature, one almost wonders if a generalized proximity relation may be defined for each prefix which can possibly be attached to the word 'proximity'! About 1963, both Pervin [84] and Leader [57] independently studied generalizations of Efremovič's original set of axioms. Pervin neglected the symmetry condition, obtaining what he called a quasi-proximity space. As well as omitting the symmetry condition, Leader used a weakened form of the 'Strong Axiom' to arrive at his topological d-space. It was shown that every topological space gives rise to a generalized proximity space (X, δ) of either form by defining the binary relation δ as follows: $A \, \delta \, B$ iff $A \cap \bar{B} \neq \varnothing$. Conversely, every quasi-proximity or topological d-space (X, δ) becomes a topological space if the closure operator is defined by $\mathrm{Cl}(A) = \{x : \{x\} \delta A\}$. Lodato [63] later added symmetry to Leader's set of axioms to obtain a symmetric binary relation which we shall refer to as a Lodato proximity. He proved that every set with a Lodato proximity defined on it satisfies the R_0 axiom (i.e. every open set contains the closure of each of its points), and that given any R_0-space we obtain a Lodato proximity compatible with the given topology if we define $A \, \delta \, B$ iff $\bar{A} \cap \bar{B} \neq \varnothing$. Mozzochi [72] has since introduced the idea of a symmetric generalized uniform structure and has studied its relationship to a Lodato proximity structure, as well as extending results of Alfsen–Fenstad, Hursch and others to such a setting.

In 1964, Hayashi [32] introduced the notion of 'paraproximity' by replacing the word 'finite' by 'arbitrary', and thereby strengthening Efremovič's 'union' axiom to read : for an arbitrary index set Λ, $(\bigcup_{\lambda \in \Lambda} A_\lambda) \delta B$ iff $A_\mu \delta B$ for some $\mu \in \Lambda$. He showed that a paraproximity space X is completely normal if one defines G to be an open set if and only if $G \, \delta \, (X - G)$. A completely normal space becomes a paraproximity space if we define $A \, \delta \, B$ iff $A \cap \bar{B} \neq \varnothing$. Hayashi [33] also discussed a generalized proximity space, which he called a pseudo-proximity space, with even weaker axioms than those considered by Pervin or Leader. He proved that every such space can be topologized and that every topological space admits a pseudo-proximity.

Recently, Leader [59] has defined a local proximity space, in which both 'proximity' and 'boundedness' are taken as primitive terms. The proximity spaces of Efremovič are the special cases in which all subsets are bounded. Just as every proximity space can be embedded as a dense subset of a compact Hausdorff space, it is shown that every local proximity space can be embedded as a dense subset of a locally compact Hausdorff space. Leader also showed that every proximity space (X, δ) with its proximity relation localized with respect to any free regular filter from (X, δ) gives rise to a local proximity space. Conversely, every local proximity space arises from the localization of some proximity relation.

CHAPTER 1

BASIC PROPERTIES

1. Introduction

In a topological space X, the topology is determined by the closure axioms given by Kuratowski concerning the relation 'x is a closure point of $A \subset X$'. When x is a closure point of A, we may say that 'x is *near A*'. In terms of this nearness relation, a continuous function $f\colon X \to Y$ may then be described as one exhibiting the property: if x is near A, then $f(x)$ is near $f(A)$. This suggests axiomatizing the relation 'A is near B' for subsets A and B of X. For the case in which X is a pseudo-metric space with pseudo-metric d, this nearness relation can be defined in a natural way. Let

$$D(A, B) = \inf\{d(a,b) : a \in A, b \in B\}.$$

We may then define:

A is near B if and only if $D(A, B) = 0$.

In terms of D, the closure of a set A is $\bar{A} = \{x : D(A, x) = 0\}$. But the nearness relation so defined goes a little further. Let (Y, e) be another pseudo-metric space, E be defined in a similar manner to D, and f be a function from X to Y. Then f is uniformly continuous if and only if $D(A, B) = 0$ implies $E(f(A), f(B)) = 0$ (see (4.8)). Thus the nearness relation between the subsets is somehow connected with uniformity.

The above nearness relation (in a pseudo-metric space) satisfies the following properties, where we denote 'A is near B' by $A \, \delta \, B$:

(1.1) $A \, \delta \, B$ implies $B \, \delta \, A$.

(1.2) $(A \cup B) \delta C$ iff $A \, \delta \, C$ or $B \, \delta \, C$.

(1.3) $A \, \delta \, B$ implies $A \neq \varnothing,\ B \neq \varnothing$.

(1.4) $A \, \delta \, B$ implies there exists a subset E such that $A \, \delta \, E$
$\qquad\qquad\qquad\qquad\qquad$ and $(X - E) \delta B$.

(1.5) $A \cap B \neq \varnothing$ implies $A \, \delta \, B$.

[7]

In a metric space, the nearness relation also satisfies:

(1.6) $x \,\delta\, y$ implies $x = y$.

(Strictly speaking one should use the notation $\{x\}\,\delta\,\{y\}$, but we shall simply write $x\,\delta\,y$.)

All of the above properties, except perhaps (1.4), are immediate consequences of the definition of the nearness relation. To verify (1.4), we note that if $A \,\delta\!\!\!/\, B$ then $D(A, B) = \epsilon > 0$. Setting

$$E = \{x \in X : D(x, B) \leqslant \epsilon/2\} \quad \text{we obtain} \quad D(A, E) \geqslant \epsilon/2$$

and $D(X - E, B) \geqslant \epsilon/2$, from which the desired property follows.

The above discussion leads to the following definition of a proximity space:

(1.7) DEFINITION. *A binary relation δ on the power set of X is called an (Efremovič) proximity on X iff δ satisfies the axioms (1.1)–(1.5). The pair (X, δ) is called a* proximity space.

Proximity relations satisfying Axiom (1.6) will be referred to as *separated* (or *Hausdorff*) proximity relations. If a proximity is derived from a (pseudo-) metric, then it is called a *(pseudo-) metric proximity*.

(1.8) REMARKS. The above axioms are different from, although equivalent to, the original axioms of Efremovič. The reason for writing the axioms in this way is to permit a smooth transition to generalized proximity spaces. It will be shown presently that a proximity δ on X induces a *topology* $\tau = \tau(\delta)$ on X if one defines the closure $\bar{A}$ of A to be the set $\{x : x\,\delta\,A\}$. It will be seen that this topology is always completely regular: in fact it is always Tychonoff if Axiom (1.6), which is equivalent to the T_1-axiom, is satisfied. Conversely, if (X, τ) is any completely regular topological space, then there exists a proximity δ on X such that $\tau(\delta) = \tau$.

Actually it would be sufficient to concern ourselves solely with separated proximity spaces, as many authors do; for if a given space fails to satisfy condition (1.6), we can instead consider the separated quotient space formed from the equivalence classes consisting of all points near to one another. However, for the sake of generality, we shall not assume a proximity space to be separated.

Suppose δ' is a binary relation on the power set of X that satisfies (1.2)–(1.5) and

(1.2′) $A\,\delta'\,(B\cup C)$ iff $A\,\delta'\,B$ or $A\,\delta'\,C$.

Then δ' induces a topology $\tau(\delta')$ on X if one defines the closure $\bar{A}$ of A to be the set $\{x : x\,\delta'\,A\}$. If in addition we require that δ' satisfy the Symmetry Axiom (1.1), then the induced topology will be completely regular. Thus we see that the Symmetry Axiom (1.1) is in a sense equivalent to the complete regularity of the induced topology.

Axiom (1.4) plays an important role in the theory of proximity spaces, but is omitted or replaced by a weaker condition in some generalized proximity spaces. It will, therefore, be helpful to avoid the use of this axiom as far as possible. We shall refer to this axiom as 'the Strong Axiom.' It should be noted that the order of the sets in the Strong Axiom is important, particularly in generalized proximity spaces in which (1.1) is not satisfied. We shall strictly observe the order even in proximity spaces, so that the proofs can then be carried over to more general situations.

Given below is an outline of a number of examples of proximity spaces in which the proximity is not constructed from a pseudometric. Since some of these will be carefully taken up in later sections, the details are not verified here.

(1.9) EXAMPLE. Just as discrete and indiscrete topologies can be defined on any set, we have discrete and indiscrete proximities. If we define $A\,\delta_1\,B$ iff $A\cap B \neq \varnothing$, then δ_1 is the *discrete proximity* on X. On the other hand, if $A\,\delta_2\,B$ for every pair of non-empty subsets A and B of X, then we obtain the *indiscrete proximity* on X.

(1.10) EXAMPLE. Given a completely regular space (X,τ), we say that subsets A and B of X are *functionally distinguishable* iff there is a continuous function $f\colon X \to [0,1]$ such that $f(A) = 0$ and $f(B) = 1$.

We may then define a proximity δ on X by

(1.11) $A\,\delta\!\!\!/\,B$ iff A and B are functionally distinguishable.

For details, refer to Theorem (2.10) and Remarks (3.15).

(1.12) EXAMPLE. Given a uniform space $(X, \mathscr{U})$, one may define a proximity on X by $A \, \delta \, B$ iff for every $U \in \mathscr{U}$, one (and hence all) of the three following equivalent conditions is satisfied:

  (i) $U[A] \cap B \neq \varnothing$;
  (ii) $A \cap U[B] \neq \varnothing$;
  (iii) $(A \times B) \cap U \neq \varnothing$.

Several interesting results concerning the relationship between uniformities and proximities will be discussed in Chapter 3.

(1.13) EXAMPLE. If $(X, \cdot, \tau)$ is a topological group and $\mathscr{N}$ is the neighbourhood system of the identity, we may define

$$A \, \delta_1 \, B \quad \text{iff for every} \quad N \in \mathscr{N}, \quad NA \cap B \neq \varnothing.$$

A second proximity δ_2 may be defined by

$$A \, \delta_2 B \quad \text{iff for every} \quad N \in \mathscr{N}, \quad AN \cap B \neq \varnothing.$$

In general the two proximities δ_1 and δ_2 differ. They coincide, however, if X is either commutative or compact.

2. Topology induced by a proximity

In this section we consider the topology on X which is induced by a proximity on X, and study its elementary properties. Properties (i) and (ii) of the following lemma, which follow directly from Axioms (1.1), (1.2) and (1.4), are useful in several proofs.

(2.1) LEMMA. (i) *If $A \, \delta \, B, A \subset C$ and $B \subset D$, then $C \, \delta \, D$. Hence X is near every non-empty subset.*

               (ii) *If there exists an x such that $A \, \delta \, x$ and $x \, \delta \, B$, then $A \, \delta \, B$.*

Note that the Strong Axiom is not used in the proof of the following theorem.

(2.2) THEOREM. *If a subset A of a proximity space (X, δ) is defined to be closed iff $x \, \delta \, A$ implies $x \in A$, then the collection of complements of all closed sets so defined yields a topology $\tau = \tau(\delta)$ on X.*

Proof: Obviously $\varnothing$ and X are closed sets. Let $\{A_i : i \in I\}$ be an arbitrary collection of closed sets. If $x \, \delta \bigcap\limits_{i \in I} A_i$ then by Lemma

(2.1), $x \delta A_i$ for each $i \in I$, and so $x \in A_i$ for each $i \in I$ since each A_i is closed. Thus $x \in \bigcap_{i \in I} A_i$, which means $\bigcap_{i \in I} A_i$ is closed. Finally, if A_1 and A_2 are closed and $x \delta (A_1 \cup A_2)$ then by (1.2), either $x \delta A_1$ or $x \delta A_2$. But A_1 and A_2 are closed, implying that $x \in A_1$ or $x \in A_2$, i.e. $x \in (A_1 \cup A_2)$. Thus $A_1 \cup A_2$ is closed.

(2.3) **THEOREM.** *Let (X, δ) be a proximity space and $\tau = \tau(\delta)$. Then the τ-closure $\bar{A}$ of a set A is given by*

$$\bar{A} = \{x : x \delta A\}.$$

Proof: If $\bar{A}$ denotes the intersection of all closed sets containing A and $A^{\delta} = \{x : x \delta A\}$, then we must show that $\bar{A} = A^{\delta}$. If $x \in A^{\delta}$ then $x \delta A$. By (2.1) this implies $x \delta \bar{A}$ and, since $\bar{A}$ is closed, $x \in \bar{A}$. Thus $A^{\delta} \subset \bar{A}$. To prove the reverse inclusion it suffices to prove that A^{δ} is closed, i.e. $x \delta A^{\delta}$ implies $x \in A^{\delta}$. Assuming $x \notin A^{\delta}$, then $x \delta A$ so that, by the Strong Axiom, there is a set E such that $x \delta E$ and $(X - E) \delta A$. Thus no point of $(X - E)$ is near A, i.e. $A^{\delta} \subset E$, which together with $x \delta E$ implies that $x \delta A^{\delta}$.

(2.4) **COROLLARY.** *If G is a subset of a proximity space (X, δ), then $G \in \tau(\delta)$ iff $x \delta (X - G)$ for every $x \in G$.*

(2.5) **COROLLARY.** *If A and B are subsets of a proximity space (X, δ), then $A \delta B$ implies*

$$\text{(i)} \quad \bar{B} \subset (X - A) \quad and \quad \text{(ii)} \quad B \subset Int\,(X - A),$$

where the closure and interior are taken with respect to $\tau(\delta)$.

Proof: Statement (i) follows directly from (2.1). To prove (ii), we use the identity: $Int\,(X - A) = X - \bar{A}$. Then $x \notin Int\,(X - A)$ implies $x \in \bar{A}$, so that $x \delta A$ and hence $x \notin B$.

(2.6) **REMARKS.** Theorem (2.3) is true if we omit the Symmetry Axiom (1.1) and add (1.2'). An alternative method of introducing the same topology on a proximity space (X, δ) would be to define for each subset A of X,

(2.7) $$A^{\delta} = \{x : x \delta A\}$$

and show that $^\delta$ is a Kuratowski closure operator as follows:

(i) By (1.3), $x \, \delta \, \varnothing$ implies $\varnothing = \varnothing^\delta$.

(ii) By (1.5), $x \in A$ implies $x \, \delta \, A$, so that $A \subset A^\delta$.

(iii) By (1.2), $x \in (A \cup B)^\delta$ iff $x \, \delta \, (A \cup B)$ iff $x \, \delta \, A$ or $x \, \delta \, B$ iff $x \in A^\delta$ or $x \in B^\delta$ iff $x \in (A^\delta \cup B^\delta)$. Thus $(A \cup B)^\delta = A^\delta \cup B^\delta$.

(iv) To prove $(A^\delta)^\delta \subset A^\delta$, suppose $x \notin A^\delta$, i.e. $x \, \bar\delta \, A$. Then by the Strong Axiom, there exists an E such that $x \, \bar\delta \, E$ and $(X - E) \, \bar\delta \, A$. Now $A^\delta \subset E$ and $x \, \bar\delta \, E$, so that $x \, \bar\delta \, A^\delta$ and $x \notin (A^\delta)^\delta$.

(2.8) LEMMA. *For subsets A and B of a proximity space (X, δ),*

$$A \, \delta \, B \quad iff \quad \bar A \, \delta \, \bar B,$$

where the closure is taken with respect to $\tau(\delta)$.

Proof: Necessity is a trivial consequence of (2.1). To prove sufficiency, suppose $A \, \bar\delta \, B$. Then by the Strong Axiom, there exists an E such that $A \, \bar\delta \, E$ and $(X - E) \, \bar\delta \, B$. By Corollary (2.5), $\bar B \subset E$ and by (2.1), $A \, \bar\delta \, E$ implies $A \, \bar\delta \, \bar B$. It then follows from the Symmetry Axiom that $\bar A \, \bar\delta \, \bar B$.

(2.9) DEFINITION. *If on a set X there is a topology τ and a proximity δ such that $\tau = \tau(\delta)$, then τ and δ are said to be* compatible.

(2.10) THEOREM. *If (X, τ) is a completely regular space, then the proximity δ defined by (1.11), namely $A \, \delta \, B$ iff A and B are functionally distinguishable, is compatible with τ. If (X, τ) is a Tychonoff space, then δ is separated.*

Proof: To show that δ is a proximity it suffices to prove that the Strong Axiom is satisfied, since the other proximity axioms are easily verified. Suppose $A \, \bar\delta \, B$ and let f be a continuous function from X to $[0, 1]$ such that $f(A) = 0$ and $f(B) = 1$. Set

$$E = \{x \in X : 1/2 \leqslant f(x) \leqslant 1\}.$$

Then $A \, \bar\delta \, E$ and $(X - E) \, \bar\delta \, B$. For instance, if g denotes the self-mapping of $[0, 1]$ defined by

$$g(y) = 2y \quad 0 \leqslant y \leqslant 1/2,$$
$$= 1 \quad 1/2 \leqslant y \leqslant 1$$

then g (and hence gf) is continuous, and $gf : X \to [0, 1]$ is such that $gf(A) = 0$ and $gf(E) = 1$.

To see that δ is separated if (X, τ) is Tychonoff, we note that if $x \neq y$ then $x \notin \bar{y}$ since (X, τ) is T_1. From the definition of a completely regular space, we are assured that x and $\bar{y}$ are functionally distinguishable, implying that $x \,\overline{\delta}\, y$.

We now show that $\tau = \tau(\delta)$. Let $G \in \tau$ and $x \in G$. Then x is not in the closed set $X - G$, so that there exists a continuous function $f: X \to [0, 1]$ such that $f(x) = 0$ and $f(X - G) = 1$, i.e. $x \,\overline{\delta}\, (X - G)$. Hence by (2.4), $G \in \tau(\delta)$. Conversely, if $G \in \tau(\delta)$ and $x \in G$, then $x \,\overline{\delta}\, (X - G)$. Hence there exists a τ-continuous function

$$f: X \to [0, 1]$$

such that $f(x) = 0$ and $f(X - G) = 1$. Then $f^{-1}([0, 1/2))$ is a τ-open neighbourhood of x contained in G. We therefore have $G \in \tau$.

Thus we know that on every completely regular (Tychonoff) space, we can define a compatible (separated) proximity. The converse of Theorem (2.10) is also true; that is, the topology $\tau(\delta)$ induced by a (separated) proximity δ is always (Tychonoff) completely regular (see Theorem (3.14)).

We now show that in a T_4 (normal $+ T_1$) space, there is another way of defining a compatible proximity.

(2.11) THEOREM. *In a T_4-space (X, τ).*

$$(2.12) \qquad A \,\delta\, B \quad iff \quad \bar{A} \cap \bar{B} \neq \varnothing$$

defines a compatible proximity.

Proof: That (2.12) defines a proximity follows from Theorem (2.10) and the fact that, in a normal space, $\bar{A} \cap \bar{B} = \varnothing$ iff $\bar{A}$ and $\bar{B}$ are functionally distinguishable. (Note that for this part, the T_1-axiom is unnecessary.) To show that $\tau = \tau(\delta)$, we observe that a T_4-space is Tychonoff and apply Theorem (2.10).

(2.13) REMARK. In a normal space (X, τ), the proximities defined by (1.11) and by (2.12) are equivalent.

(2.14) THEOREM. *If a completely regular space (X, τ) has a compatible proximity δ defined by (2.12), then X is normal.*

Proof: If A and B are disjoint closed sets, then $A \,\overline{\delta}\, B$. By the Strong Axiom, there exists an E such that $A \,\overline{\delta}\, E$ and $(X - E) \,\overline{\delta}\, B$.

From 2.5 (ii), we then have $A \subset \text{Int}(X - E)$ and $B \subset \text{Int} E$. Since $\text{Int} E \cap \text{Int}(X - E) = \varnothing$, X is normal.

Just as the class of topologies on a given set can be partially ordered by inclusion, one can impose a partial order on the class of proximities defined on a set in the following manner:

(2.15) DEFINITION. *If δ_1 and δ_2 are two proximities on a set X, we define*

(2.16) $\qquad \delta_1 > \delta_2 \quad iff \quad A \, \delta_1 B \quad implies \quad A \, \delta_2 B.$

The above is expressed by saying that δ_1 is *finer* than δ_2, or δ_2 is *coarser* than δ_1.

The following theorem shows that a finer proximity structure induces a finer topology:

(2.17) THEOREM. (*a*) *Let δ_1 and δ_2 be two proximities defined on a set X. Then $\delta_1 < \delta_2$ implies $\tau(\delta_1) \subset \tau(\delta_2)$.*

(*b*) *Let τ_1 and τ_2 be two completely regular topologies on X, and let δ_1 and δ_2 be the proximities on X defined by* (1.11) *with respect to τ_1 and τ_2 respectively. Then $\tau_1 \subset \tau_2$ implies $\delta_1 < \delta_2$.*

Proof: (*a*) Suppose $A \in \tau(\delta_1)$. Then by (2.4) $x \, \delta_1 (X - A)$ for each $x \in A$. Moreover, since $\delta_1 < \delta_2$, $x \, \delta_2 (X - A)$ for each $x \in A$. Thus $A \in \tau(\delta_2)$, from which we conclude that $\tau(\delta_1) \subset \tau(\delta_2)$.

(*b*) If $A \, \delta_1 B$, then there exists a τ_1-continuous function $f: (X, \tau_1) \to [0, 1]$ such that $f(A) = 0$ and $f(B) = 1$. Since $\tau_1 \subset \tau_2$, f is also a τ_2-continuous function, showing that $A \, \delta_2 B$. Hence $\delta_1 < \delta_2$.

(2.18) REMARKS. For a slightly stronger form of Theorem 2.17 (*b*), the reader is referred to Remarks (3.15).

It is sometimes said that a proximity is a finer structure on a set than a topology, since a topological space may have different compatible proximities. For example, let X be the real line with the usual topology and let δ_1 be defined by $A \, \delta_1 B$ iff $D(A, B) = 0$, where $D(A, B) = \inf\{|a - b| : a \in A, b \in B\}$. Let δ_2 be defined by (2.12). Then both δ_1 and δ_2 are compatible with the topology of X. However, the sets $A = \{n : n \in N\}$ and $B = \{n - 1/n : n \in N\}$ are such that $A \, \delta_1 B$ but $A \, \delta_2 B$.

3. Alternate description of proximity

Given a uniform space $(X, \mathscr{U})$, a subset B may be said to be a uniform neighbourhood of A iff there is an entourage $U \in \mathscr{U}$ such that $U[A] \subset B$. An analogous concept, that of a δ-neighbourhood, can be introduced in a proximity space and furnishes an alternative approach to the study of proximity spaces. This concept of a δ-neighbourhood is not only useful in the theory of proximity spaces, but is also, in a sense, dual to the concept of proximity. This duality will be studied in detail in Section 6.

(3.1) DEFINITION. *A subset B of a proximity space (X, δ) is a δ-neighbourhood of A (in symbols $A \ll B$) iff $A \mathbin{\bar\delta} (X - B)$.*

The second part of the following lemma (which is a strengthened form of (2.5)) justifies the term 'δ-neighbourhood'.

(3.2) LEMMA. *Let (X, δ) be a proximity space and let $\bar{A}$ and $\operatorname{Int} A$ denote, respectively, the closure and interior of A in $\tau(\delta)$. Then*

(i) *$A \ll B$ implies $\bar{A} \ll B$, and*
(ii) *$A \ll B$ implies $A \ll \operatorname{Int} B$.*

Therefore $A \subset \operatorname{Int} B$, showing that a δ-neighbourhood is a topological neighbourhood.

Proof: (i) Using (2.8), $A \mathbin{\bar\delta} (X - B)$ implies $\bar{A} \mathbin{\bar\delta} (X - B)$, i.e. $\bar{A} \ll B$.

(ii) $A \mathbin{\bar\delta} (X - B)$ implies $A \mathbin{\bar\delta} (\overline{X - B})$. Equivalently,

$$A \mathbin{\bar\delta} (X - \operatorname{Int} B), \quad \text{i.e.} \quad A \ll \operatorname{Int} B.$$

The following is an alternate way of stating the Strong Axiom and is employed by many authors:

(3.3) LEMMA. *The Strong Axiom (1.4) is equivalent to*

(3.4) *$A \mathbin{\bar\delta} B$ implies there exist subsets C and D such that*

$$A \mathbin{\bar\delta} (X - C), \quad (X - D) \mathbin{\bar\delta} B \quad \text{and} \quad C \mathbin{\bar\delta} D.$$

Proof: To prove that (3.4) implies (1.4), we note that if $C \mathbin{\bar\delta} D$, then $C \subset (X - D)$. Setting $E = X - C$, we have $A \mathbin{\bar\delta} E$ and $(X - E) \mathbin{\bar\delta} B$. On the other hand, suppose (1.4) holds. Then

$A \not\delta B$ implies there is a D such tht $A \not\delta D$ and $(X - D) \not\delta B$. Moreover, there exists a C such that $A \not\delta (X - C)$ and $C \not\delta D$, completing the proof.

(3.5) COROLLARY. *$A \not\delta B$ implies there exist subsets C and D such that $A \ll C$, $B \ll D$ and $C \not\delta D$. Consequently, the topology $\tau(\delta)$ of a separated proximity space (X, δ) is Hausdorff.*

(3.6) LEMMA. *Let δ be a compatible proximity on a completely regular space (X, τ). If A is compact, B is closed and $A \cap B = \varnothing$, then $A \not\delta B$.*

Proof: For each $a \in A$, $a \not\delta B$. From (3.2) and (3.5), we know there exists an open neighbourhood N_a of a such that $N_a \not\delta B$. Now $\{N_a : a \in A\}$ is an open cover of the compact set A, and so there is a finite subcover $\{N_{a_i} : i = 1, ..., n\}$. By (1.2), $N \not\delta B$, where $N = \bigcup\limits_{i=1}^{n} N_{a_i}$. But $A \subset N$, implying that $A \not\delta B$.

It is well known that a compact completely regular space has a unique compatible uniformity. A similar theorem is true in proximity spaces:

(3.7) THEOREM. *Every compact space which is completely regular (Tychonoff) has a unique compatible (separated) proximity, given by*
$$A \not\delta B \quad iff \quad \bar{A} \cap \bar{B} \neq \varnothing.$$

Proof: If δ is any compatible proximity and $\bar{A} \cap \bar{B} \neq \varnothing$, then by (1.5) and (2.8), $A \not\delta B$. Since closed subsets of a compact space are compact, the foregoing lemma implies the converse.

(3.8) REMARKS. Contrary to a conjecture made by Alexandroff, there also exist non-compact completely regular spaces with unique compatible proximities. (Such spaces are necessarily normal, as will be seen in Section 7.) The following is one such example: the space of all ordinals less than the first uncountable ordinal with the order topology is not compact, but has a unique compatible proximity.

In a later section (Theorem (7.20)), we shall actually characterize those topological spaces which possess unique compatible proximities.

(3.9) THEOREM. *Given a proximity space* (X, δ), *the relation* $\ll$ *satisfies the following properties*:

(i) $X \ll X$.

(ii) $A \ll B$ *implies* $A \subset B$. *The converse holds iff* (X, δ) *is discrete*.

(iii) $A \subset B \ll C \subset D$ *implies* $A \ll D$.

(iv) $A \ll B_i$ *for* $i \ldots 1, \ldots, n$ *iff* $A \ll \bigcap_{i=1}^{n} B_i$.

(v) $A \ll B$ *implies* $(X - B) \ll (X - A)$.

(vi) $A \ll B$ *implies there is a* C *such that* $A \ll C \ll B$.

If δ is a separated proximity, then

(vii) $x \ll (X - y)$ *iff* $x \neq y$.

Proof: (i) Since $X \delta \varnothing$ by (1.3), $X \ll X$.

(ii) If $A \delta (X - B)$ then $A \cap (X - B) = \varnothing$, implying $A \subset B$. The second part is an easy consequence of the definition of discrete proximity given in Example (1.9).

(iii) If $A \not\ll D$, then $A \delta (X - D)$. This implies that $B \delta (X - C)$ or $B \not\ll C$, a contradiction.

(iv) It suffices to consider $n = 2$. $A \ll B_1$ and $A \ll B_2$ iff $A \delta (X - B_1)$ and $A \delta (X - B_2)$ iff (by (1.2)) $A \delta [(X - B_1) \cup (X - B_2)]$ iff $A \delta [X - (B_1 \cap B_2)]$ iff $A \ll (B_1 \cap B_2)$.

(v) $A \ll B$ implies $A \delta (X - B)$. By (1.1), $(X - B) \delta A$, i.e. $(X - B) \ll (X - A)$.

(vi) $A \ll B$ implies $A \delta (X - B)$. By the Strong Axiom, there exists a set $(X - C)$ such that $A \delta (X - C)$ and $C \delta (X - B)$; that is, $A \ll C \ll B$.

(vii) $x \neq y$ iff $x \delta y$ (by (1.6)) iff $x \ll (X - y)$.

(3.10) COROLLARY. $A_i \ll B_i$ *for* $i = 1, \ldots, n$ *implies*

$$\bigcap_{i=1}^{n} A_i \ll \bigcap_{i=1}^{n} B_i \quad and \quad \bigcup_{i=1}^{n} A_i \ll \bigcup_{i=1}^{n} B_i.$$

All of the separated proximity axioms (1.1)–(1.6) are used in the above proofs. In particular we note that (1.4) is equivalent to 3.9 (vi), and (1.6) is equivalent to 3.9 (vii). The following is a converse of Theorem (3.9).

(3.11) THEOREM. *If $\ll$ is a binary relation on the power set of X satisfying 3.9 (i)–(vi) and δ is defined by*

(3.12) $A \, \delta \, B \quad iff \quad A \ll (X - B),$

then δ is a proximity on X. B is a δ-neighbourhood of A iff $A \ll B$. Moreover, if $\ll$ also satisfies 3.9 (vii), then δ is separated.

Proof: (i) $A \, \delta \, B$ implies $A \ll (X - B)$. By 3.9(v), $B \ll (X - A)$, and so $B \, \delta \, A$.

(ii) $(A \cup B) \, \delta \, C$ implies $(A \cup B) \ll (X - C)$. Then by 3.9(iii), $A \ll (X - C)$ and $B \ll (X - C)$; that is, $A \, \delta \, C$ and $B \, \delta \, C$. Conversely, if $(A \cup B) \, \delta \, C$ then by part (i), $C \, \delta \, (A \cup B)$. Hence $C \not\ll [X - (A \cup B)]$, or $C \not\ll [(X - A) \cap (X - B)]$. Thus by 3.9(iv), $C \not\ll (X - A)$ or $C \not\ll (X - B)$. Hence $C \, \delta \, A$ or $C \, \delta \, B$ and it follows, since δ is symmetric, that $A \, \delta \, C$ or $B \, \delta \, C$.

(iii) (1.3) is a direct consequence of 3.9(i).

(iv) Suppose $A \, \delta \, B$, i.e. $A \ll (X - B)$. Then 3.9(vi) assures the existence of a C such that $A \ll (X - C) \ll (X - B)$. Thus there is a C such that $A \, \delta \, C$ and $(X - C) \, \delta \, B$.

(v) If $A \, \delta \, B$, then $A \ll (X - B)$. From 3.9(ii) we have

$$A \subset (X - B), \quad \text{i.e.} \quad A \cap B = \varnothing.$$

If 3.9 (vii) is satisfied, (1.6) follows immediately. That B is a δ-neighbourhood of A iff $A \ll B$ follows easily from the definitions of the terms involved.

In a uniform space $(X, \mathscr{U})$, the closure of any subset A is given by

$$\bar{A} = \bigcap_{U \in \mathscr{U}} U[A].$$

The following result is an analogue of this in proximity spaces.

(3.13) THEOREM. *If (X, δ) is a proximity space and $A \subset X$, then*

$$\bar{A} = \bigcap_{A \ll B} B.$$

Proof: From 3.2(i) and 3.9(ii) we conclude that $A \ll B$ implies $\bar{A} \subset B$, and hence $\bar{A} \subset \bigcap_{A \ll B} B$. To show the reverse inclusion, suppose that $x \notin \bar{A}$. Then $x \, \delta \, \bar{A}$ and, by (3.5), $\bar{A}$ has a δ-neighbourhood B_x not containing x. Thus $x \notin \bigcap_{A \ll B} B$.

In Theorem (2.10) it was shown that every completely regular space (X, τ) has a compatible proximity δ. The following is a converse.

(3.14) THEOREM. *If (X, δ) is a (separated) proximity space, then $\tau(\delta)$ is (Tychonoff) completely regular.*

Proof: That $\tau(\delta)$ is T_1 if δ is separated follows easily from Axiom (1.6). We now indicate briefly why $\tau(\delta)$ is completely regular. If A is a closed set and $x \notin A$, then $x \not\delta A$. Hence we have $x \ll (X - A)$ and, after applying 3.9(vi) twice, we find that there are sets B and C such that $x \ll B \ll C \ll (X - A)$. Using Lemma (3.2), we obtain $x \subset \text{Int}\, B \subset \bar{B} \subset \text{Int}\, C \subset \bar{C} \subset (X - A)$. This is similar to the main step in the proof of Urysohn's Lemma, and proceeding as therein, we obtain a continuous function $f: X \to [0, 1]$ such that $f(x) = 0$ and $f(A) = 1$.

(3.15) REMARKS. Actually, following the proof of Urysohn's Lemma one can prove that for a compatible proximity δ defined on X, $A \not\delta B$ implies the existence of a continuous function $f: X \to [0, 1]$ such that $f(A) = 0$ and $f(B) = 1$ (see Theorem (7.12)). Consequently, the proximity defined by (1.11), namely $A \not\delta B$ iff A and B are functionally distinguishable, is the largest or finest compatible proximity which can be defined on a completely regular space. Or, as Smirnov proved: every completely regular space has a maximal associated proximity space. It is useful to note that, as a result of Corollary (2.13), the proximity defined by (2.12) is the largest compatible proximity that can be defined on a normal T_1-space.

We can now strengthen 2.17(b) to read: Let τ_1 and τ_2 be two completely regular topologies on X, δ_1 be any proximity on X compatible with τ_1, and δ_2 be defined by (1.11) with respect to τ_2. Then $\tau_1 \subset \tau_2$ implies $\delta_1 < \delta_2$.

4. Subspaces and products of proximity spaces

In the study of general topological spaces, continuous functions play an important role. A similar role is played by uniformly continuous functions in uniform spaces. Their analogue in the theory of proximity spaces is the concept of a proximity (or *proximally continuous* or *equicontinuous* or δ-) mapping.

(4.1) DEFINITION. *Let* (X, δ_1) *and* (Y, δ_2) *be two proximity spaces. A function* $f: X \to Y$ *is said to be a* proximity mapping *iff*

$$A \, \delta_1 B \quad \text{implies} \quad f(A) \, \delta_2 f(B).$$

Equivalently, f is a proximity mapping iff

$$C \, \delta_2 D \quad \text{implies} \quad f^{-1}(C) \, \delta_1 f^{-1}(D),$$

or $\qquad\qquad C \ll_2 D \quad \text{implies} \quad f^{-1}(C) \ll_1 f^{-1}(D).$

It is easy to see that the composition of two proximity mappings is a proximity mapping. The next theorem is similar to the well-known result: a uniformly continuous function is continuous with respect to the induced topologies.

(4.2) THEOREM. *A proximity mapping* $f: (X, \delta_1) \to (Y, \delta_2)$ *is continuous with respect to* $\tau(\delta_1)$ *and* $\tau(\delta_2)$.

Proof: This result follows easily from the fact that $x \, \delta_1 A$ implies $f(x) \, \delta_2 f(A)$, i.e. $f(\bar{A}) \subset \overline{f(A)}$.

(4.3) REMARKS. The converse of the foregoing theorem is false. Consider the example of (2.18): the identity mapping on X is continuous with respect to $\tau(\delta_1)$ and $\tau(\delta_2)$, but is not a proximity mapping from (X, δ_1) to (X, δ_2).

It is natural to inquire as to when the converse of Theorem (4.2) is true. Recalling that a continuous function on a compact space is uniformly continuous, we observe from the next theorem that a completely analogous result holds in proximity spaces. However, compactness is too strong and we shall prove a better result when we consider equinormal proximity spaces (see Theorem (7.22)).

(4.4) THEOREM. *If* (X, δ_1) *and* (Y, δ_2) *are proximity spaces and* X *is compact, then every continuous function* f *from* X *to* Y *is a proximity mapping.*

Proof: If A and B are subsets of X such that $A \, \delta_1 B$, then $\bar{A} \cap \bar{B} \neq \varnothing$ by Theorem (3.7). But this implies that

$$f(\bar{A}) \cap f(\bar{B}) \neq \varnothing, \quad \text{i.e.} \quad f(\bar{A}) \, \delta_2 f(\bar{B}).$$

Since f is continuous, $f(\bar{A}) \subset \overline{f(A)}$ and $f(\bar{B}) \subset \overline{f(B)}$, yielding $\overline{f(A)} \delta_2 \overline{f(B)}$. From Lemma (2.8) it follows that $f(A) \delta_2 f(B)$, and we conclude that f is a proximity mapping.

(4.5) THEOREM. *Given a function* $f: X \to (Y, \delta_2)$, *the coarsest proximity* δ_0 *which may be assigned to* X *in order that* f *be proximally continuous is defined by*

(4.6) $A \delta_0 B$ *iff there exists a* $C \subset Y$ *such that*

$$f(A) \delta_2 (Y - C) \quad and \quad f^{-1}(C) \subset (X - B).$$

Proof: We first verify that δ_0 is a proximity on X.

(i) Suppose $A \not\!\delta_0 B$ and let $D = (Y - f(A))$. Since

$$f(B) \subset (Y - C) \quad and \quad f(A) \not\!\delta_2 (Y - C),$$

we have $f(B) \not\!\delta_2 (Y - D)$. Moreover,

$$f^{-1}(D) = X - f^{-1}(f(A)) \subset (X - A).$$

Hence $B \not\!\delta_0 A$.

(ii) $(A \cup B) \delta_0 C$ implies the existence of a $D \subset Y$ such that $[f(A) \cup f(B)] \delta_2 (Y - D)$ and $f^{-1}(D) \subset (X - C)$, from which $A \delta_0 C$ and $B \delta_0 C$ follow. If $A \not\!\delta_0 C$ and $B \not\!\delta_0 C$, there exist D_1 and D_2 such that $f(A) \not\!\delta_2 (Y - D_1)$, $f(B) \not\!\delta_2 (Y - D_2)$, $f^{-1}(D_1) \subset (X - C)$ and $f^{-1}(D_2) \subset (X - C)$. Therefore $[f(A) \cup f(B)] \not\!\delta_2 [Y - (D_1 \cup D_2)]$ and $f^{-1}(D_1 \cup D_2) \subset (X - C)$, i.e. $(A \cup B) \not\!\delta_0 C$.

(iii) If $A = \varnothing$, then $f(A) \not\!\delta_2 Y$ and $f^{-1}(\varnothing) \subset (X - B)$. Hence $A \not\!\delta_0 B$.

(iv) If $A \not\!\delta_0 B$, then there exists a $C \subset Y$ such that

$$f^{-1}(C) \subset (X - B) \quad and \quad f(A) \not\!\delta_2 (Y - C).$$

This latter relation and (1.4) together assure the existence of a set D such that $f(A) \not\!\delta_2 D$ and $(Y - D) \not\!\delta_2 (Y - C)$. Let $E = f^{-1}(D)$. Since $f(A) \not\!\delta_2 D$, $A \not\!\delta_0 E$. As $f(X - E) \subset (Y - D) \not\!\delta_2 (Y - C)$ and $f^{-1}(C) \subset (X - B)$, we have $(X - E) \not\!\delta_0 B$.

(v) $A \not\!\delta_0 B$ implies there exists a $C \subset Y$ such that

$$f(A) \not\!\delta_2 (Y - C) \quad and \quad f^{-1}(C) \subset (X - B).$$

Therefore $f(A) \cap (Y - C) = \varnothing$ and $f^{-1}(f(A)) \cap f^{-1}(Y - C) = \varnothing$. Since $A \subset f^{-1}(f(A))$ and $B \subset f^{-1}(Y - C)$, we have $A \cap B = \varnothing$.

In order to show that $f: (X, \delta_0) \to (Y, \delta_2)$ is proximally continuous, suppose that $f(A)\, \delta_2\, f(B)$. Since $f(A) \ll (Y - f(B))$, there exists a C such that $f(A) \ll C \ll (Y - f(B))$ by 3.9(vi). Thus $f(A)\, \delta_2\, (Y - C)$ and $f^{-1}(C) \subset (X - f^{-1}(f(B))) \subset (X - B)$, i.e. $A\, \delta_0\, B$.

It remains to show that if δ_1 is any proximity on X such that $f: (X, \delta_1) \to (Y, \delta_2)$ is proximally continuous, then δ_1 is finer than δ_0. If $A\, \delta_0\, B$, then there exists a $C \subset Y$ such that $f(A)\, \delta_2\, (Y - C)$ and $f^{-1}(C) \subset (X - B)$. Since f is proximally continuous,

$$A\, \delta_1\, (X - f^{-1}(C)), \quad \text{and} \quad B \subset (X - f^{-1}(C))$$

implies $A\, \delta_1\, B$. Thus $\delta_1 > \delta_0$.

(4.7) COROLLARY. *If $f: (X, \delta_0) \to (Y, \delta_2)$ where δ_0 is defined by* (4.6), *then $f^{-1}[\tau(\delta_2)] \subset \tau(\delta_0)$.*

(4.8) REMARK. Perhaps we should mention at this point that in Chapter 3, it will be shown that every uniformly continuous function is proximally continuous with respect to the induced proximities.

Moreover, a function f mapping a metric space (X, d) to a metric space (Y, e) is uniformly continuous if and only if it is proximally continuous, (a result due to Efremovič). We shall simply outline a proof here, as a more general result will be proved later (see (12.20)).

Necessity follows trivially. To prove the converse, suppose that f is not uniformly continuous. Then for some $\epsilon > 0$ we can find sequences (a_n), (b_n) in X such that $d(a_n, b_n) \to 0$ while

$$e(f(a_n), f(b_n)) \geqslant \epsilon.$$

If we can find an infinite index set I such that $E(A, B) > 0$, where $A = \{f(a_n) : n \in I\}$, $B = \{f(b_n) : n \in I\}$, and

$$E(A, B) = \inf\{e(a, b) : a \in A, b \in B\},$$

then it will follow that f is not proximally continuous. If

$$C = \{f(a_n) : n \in N\} \cup \{f(b_n) : n \in N\}$$

has an infinite subset of diameter at most $\epsilon/2$, then clearly we can find a subset consisting solely of $f(a_n)$'s (or $f(b_n)$'s). In the case that all elements of this subset of $f(a_n)$'s (resp. $f(b_n)$'s) are within

$\epsilon/2$ distance of each other, every $f(b_n)$ (resp. $f(a_n)$) is at least $\epsilon/2$ distant from all $f(a_n)$'s (resp. $f(b_n)$'s) and the result follows. Otherwise, the $\epsilon/4$-neighbourhood of any of these points contains only finitely many others, and so by induction we can find an infinite set I such that $E(A, B) \geqslant \epsilon/4$, completing the proof.

(4.9) DEFINITION. *Two proximity spaces* (X, δ_1) *and* (Y, δ_2) *are called* proximally isomorphic (*or* δ-homeomorphic) *iff there exists a one-to-one mapping f from X onto Y such that both f and f^{-1} are proximity mappings. Such a mapping f is termed a* proximity isomorphism *or* δ-homeomorphism. *The term* equimorphism *is also used.*

It follows from Theorem (4.2) that proximally isomorphic spaces are homeomorphic. However, as can be seen from the example of (2.18), homeomorphic spaces are not necessarily proximally isomorphic.

A property is said to be a *proximity invariant* iff it is preserved under proximity isomorphisms. That every topological invariant is a proximity invariant is clear, again from Theorem (4.2).

(4.10) DEFINITION. *If* (X, δ) *is a proximity space and* $Y \subset X$, *then subsets of Y are also subsets of X. For subsets A and B of Y, we define*

(4.11) $$A \, \delta_Y \, B \quad \text{iff} \quad A \, \delta \, B.$$

It is easily verified that δ_Y is a proximity on Y and that $\tau(\delta_Y)$ is the subspace topology induced on Y by $\tau(\delta)$. We call δ_Y the *induced* (or *subspace*) *proximity*.

We next consider the product of a family $\{(X_a, \delta_a) : a \in \Lambda\}$ of proximity spaces. Let $X = \underset{a \in \Lambda}{\times} X_a$ denote the Cartesian product of these spaces. A proximity δ can be defined on X as follows:

(4.12) Let A and B be subsets of X. Define $A \, \delta \, B$ iff for each pair of finite covers $\{A_i : i = 1, ..., m\}$ and $\{B_j : j = 1, ..., n\}$ of A and B respectively, there exists an A_i and a B_j such that

$$P_a[A_i] \, \delta_a \, P_a[B_j] \quad \text{for each} \quad a \in \Lambda.$$

(P_a denotes the projection of X onto X_a.)

(4.13) THEOREM. *The binary relation δ defined in* (4.12) *is a proximity on the product space X. It is separated iff each δ_a is separated.*

Proof: (i) Since each δ_a is symmetric, so is δ and (1.1) is satisfied.

(ii) Let A, B and C be subsets of X and suppose $A \delta C$. Since every cover of $(A \cup B)$ is a cover of A, it follows that $(A \cup B) \delta C$. Conversely, suppose $A \delta C$ and $B \delta C$. Then there are finite covers $\{A_i : i = 1, ..., m\}$ and $\{C_j : j = 1, ..., n\}$ of A and C respectively such that $P_a[A_i] \delta_a P_a[C_j]$ for some $a = s_{ij} \in \Lambda$, where $i = 1, ..., m$ and $j = 1, ..., n$. Likewise, there are finite covers

$$\{A_i : i = m+1, ..., m+p\} \quad \text{and} \quad \{D_k : k = 1, ..., q\}$$

of B and C respectively such that $P_a[A_i] \delta_a P_a[D_k]$ for some $a = t_{ik} \in \Lambda$, where $i = m+1, ..., m+p$ and $k = 1, ..., q$. Now $\{C_j \cap D_k : j = 1, ..., n; k = 1, ..., q\}$ is a cover of C and

$$\{A_i : i = 1, ..., m+p\}$$

is a cover of $A \cup B$. Since $P_a[A_i] \delta_a P_a[C_j \cap D_k]$ for $a = s_{ij}$ or $a = t_{ik}$, we conclude that $(A \cup B) \delta C$.

(iii) That $A \delta B$ implies that A and B are non-void follows easily.

(iv) If $A \delta B$, then there exist covers $\{A_i : i = 1, ..., m\}$ and $\{B_j : j = 1, ..., n\}$ of A and B respectively such that

$$P_s[A_i] \delta_s P_s[B_j] \quad \text{for some} \quad s = t_{ij} \in \Lambda,$$

where $i = 1, ..., m$ and $j = 1, ..., n$. Since each (X_a, δ_a) is a proximity space, there exist E_{ij} such that $P_s[A_i] \delta_s E_{ij}$ and

$$(X_s - E_{ij}) \delta_s B_j.$$

Set $E_j = \bigcap_{i=1}^{m} P_s^{-1}[E_{ij}]$ and $E = \bigcup_{j=1}^{n} E_j$. Since $P_s[E_j] \subset E_{ij}$, we have $P_s[A_i] \delta_s P_s[E_j]$ for $s = t_{ij}$; that is, $A \delta E$.

Let

$$D_{ij} = (X - P_s^{-1}[E_{ij}]) = P_s^{-1}[X_s - E_{ij}]$$

and

$$F_j = (X - E_j) = \bigcup_{i=1}^{m} D_{ij}.$$

Then $(X - E) = \bigcap\limits_{j=1}^{n} F_j$. Since

$$(X_s - E_{ij})\,\delta_s\,B_j \quad \text{and} \quad D_{ij} = P_s^{-1}(X_s - E_{ij}),$$

$B_j \,\bar{\delta}\, D_{ij}$ for all i and j. This implies $B_j \,\bar{\delta}\, F_j$ for all j; thus

$$(X - E)\,\bar{\delta}\,B_j \quad \text{for all} \quad j,$$

showing that $(X - E)\,\bar{\delta}\,B$.

(v) If $A \cap B \neq \varnothing$, then there exists an $x = (x_a) \in A \cap B$. For every pair of covers $\{A_i : i = 1, \ldots, m\}$ and $\{B_j : j = 1, \ldots, n\}$ of A and B respectively, there exists an A_i and a B_j containing x. Clearly $x_a \in P_a[A_i] \cap P_a[B_j]$ for all $a \in \Lambda$, which implies

$$P_a[A_i]\,\delta_a\,P_a[B_j] \quad \text{for} \quad a \in \Lambda;$$

that is $A\,\delta\,B$.

(vi) Suppose X is separated and $x_a\delta_a y_a$ for points x_a, y_a in X_a. There are two points in X, x and y say, such that $x = (x_\lambda)$ and $y = (y_\lambda)$, where $x_\lambda = y_\lambda$ for all $\lambda \in \Lambda$ except $\lambda = a$. Then $x\,\delta\,y$, implying $x = y$. That is, $x_a = y_a$, showing δ_a is separated. Conversely, $x = (x_a)\,\delta\,y = (y_a)$ implies $x_a\delta_a y_a$ for all $a \in \Lambda$. If each δ_a is separated, then $x_a = y_a$ for all $a \in \Lambda$, i.e. $x = y$.

(4.14) COROLLARY. *A mapping f from a proximity space (Y, δ_1) to $X = \underset{a \in \Lambda}{\times} X_a$ is a proximity mapping if and only if the composition $P_a \circ f : Y \to X_a$ is a proximity mapping for each projection P_a. Consequently, δ is the smallest proximity on X for which each projection P_a is proximally continuous.*

Proof: We need only prove that if each P_a of is proximally continuous, then so is f. Let A and B be subsets of Y such that $A\,\delta_1\,B$, and let $\{A_i : i = 1, \ldots, m\}$ and $\{B_j : j = 1, \ldots, n\}$ be finite covers of $f(A)$ and $f(B)$ respectively. Then for each $a \in \Lambda$,

$$P_a[f(A)]\,\delta_a\,P_a[f(B)] \quad \text{so that} \quad P_a\!\left[\bigcup_{i=1}^{m} A_i\right]\delta_a\,P_a\!\left[\bigcup_{j=1}^{n} B_j\right].$$

Hence by (1.2), $P_a[A_i]\,\delta_a\,P_a[B_j]$ for some i,j. From (4.12) we then have $f(A)\,\delta f(B)$, showing that f is proximally continuous.

The second part of the above corollary reminds one of the manner in which the Tychonoff product topology is defined on the

Cartesian product of a collection of topological spaces. Similarly, an analogous definition to that of the quotient topology is used in the case of quotient proximity. We shall not give the definition here, however, as we do not make use of this concept.

Notes

1. The axioms for a proximity space were originally given by Efremovič [18, 19], although they appeared in a slightly different but equivalent form to those presented in this section. The equivalence of the Symmetry Axiom for a proximity to the complete regularity of the associated topology has been shown by Pervin [84].

2, 3. The theorems in these sections are mainly due to Efremovič [19], as is the concept of a δ-neighbourhood. They have been collectively presented by Smirnov [97, 98] in his early survey of proximity spaces. In the same survey, Smirnov announced that every completely regular space has a maximal associated proximity structure. The example referred to in Remarks (3.8) was contributed by Pervin [86]. For a clear proof of Urysohn's lemma, the reader is referred to page 100 of Thron [115]. It should be observed that, as with most order relations, there is no general agreement on the definition of a partial order on the set of all proximities.

4. The results concerning proximity mappings were first established by Smirnov [98]. A brief discussion of most of the topics covered in this section can be found in Dowker [17]. For an account of proximity on the product of proximity spaces, see Leader [57].

CHAPTER 2

COMPACTIFICATIONS OF PROXIMITY SPACES

5. Clusters and ultrafilters

Ultrafilters play an important role in topological spaces inasmuch as such notions as convergence and compactness can be characterized in terms of ultrafilters. In this section we consider their counterparts, namely clusters, in proximity spaces. We show that ultrafilters and clusters are closely related, and use this relationship to derive several important results in the theory of proximity spaces.

It is well known that a family $\mathscr{L}$ of subsets of a non-empty set X is an *ultrafilter* if and only if the following conditions are satisfied:

(5.1)   If A and B belong to $\mathscr{L}$, then $A \cap B \neq \varnothing$.

(5.2)   If $A \cap C \neq \varnothing$ for every $C \in \mathscr{L}$, then $A \in \mathscr{L}$.

(5.3)   If $(A \cup B) \in \mathscr{L}$, then $A \in \mathscr{L}$ or $B \in \mathscr{L}$.

It is natural to expect that the collections of sets in a proximity space satisfying conditions similar to (5.1)–(5.3), with *nearness* replacing *non-empty intersection*, should be valuable in the theory of proximity spaces. This is indeed the case and we are led to the following definition:

(5.4)   DEFINITION. *A collection σ of subsets of a proximity space (X, δ) is called a* cluster *iff the following conditions are satisfied:*

(i) *If A and B belong to σ, then $A \, \delta \, B$.*
(ii) *If $A \, \delta \, C$ for every $C \in \sigma$, then $A \in \sigma$.*
(iii) *If $(A \cup B) \in \sigma$, then $A \in \sigma$ or $B \in \sigma$.*

(5.5)   REMARKS. (i) For each $x \in X$, the collection

$$\sigma_x = \{A \subset X : A \, \delta \, x\}$$

[27]

is a cluster. We call such a cluster a *point cluster* and use the above notation.

(ii) If $\{x\} \in \sigma$ for some $x \in X$, then $\sigma = \sigma_x$. If (X, δ) is separated, then no cluster can contain more than one point since that would contradict (1.6); consequently, $x \neq y$ implies $\sigma_x \neq \sigma_y$.

(iii) If σ is any cluster in X, then $X \in \sigma$ by 5.4 (ii). Hence for each subset E of X, either $E \in \sigma$ or $(X - E) \in \sigma$. Recall that an ultrafilter also has this property.

(iv) If $A \in \sigma$ and $A \subset B$, then $B \in \sigma$. This too is a property of an ultrafilter.

(v) If σ is any cluster in X, then $A \in \sigma$ iff $\bar{A} \in \sigma$. This follows from (2.8), 5.4 (ii) and 5.5 (iv).

(5.6) LEMMA. *If σ_1 and σ_2 are two clusters in a proximity space (X, δ) such that $\sigma_1 \subset \sigma_2$, then $\sigma_1 = \sigma_2$.*

Proof. If $A \in \sigma_2$, then $A \,\delta\, C$ for every $C \in \sigma_2$. Since $\sigma_1 \subset \sigma_2$, $A \,\delta\, B$ for every $B \in \sigma_1$, which shows that $A \in \sigma_1$. Thus $\sigma_2 \subset \sigma_1$.

The following lemma on ultrafilters is useful in deriving the fundamental relationship (Theorem (5.8)) between ultrafilters and clusters, and the two together are repeatedly used in many proofs.

(5.7) LEMMA. *Let $\mathscr{P}$ be a collection of subsets of X such that (i) $\varnothing \notin \mathscr{P}$, and (ii) $(A \cup B) \in \mathscr{P}$ iff $A \in \mathscr{P}$ or $B \in \mathscr{P}$. If $A_0 \in \mathscr{P}$, then there exists an ultrafilter $\mathscr{L}$ such that*

$$(a) \quad A_0 \in \mathscr{L} \quad and \quad (b) \quad \mathscr{L} \subset \mathscr{P}.$$

Proof. By Zorn's lemma, there exists a maximal collection $\mathscr{L}$ (of subsets of X) satisfying (a) and

$$(b') \qquad A_i \in \mathscr{L} \quad \text{for} \quad i = 1, \dots, n \quad \text{implies} \quad \bigcap_{i=1}^{n} A_i \in \mathscr{P}.$$

Obviously $\varnothing \notin \mathscr{L}$. If A and B belong to $\mathscr{L}$ then by (b'), $A \cap B \in \mathscr{P}$. Since $\mathscr{L}$ is maximal, we must have $A \cap B \in \mathscr{L}$. If $A \in \mathscr{L}$ and $A \subset D$, then $D \in \mathscr{P}$ and hence belongs to $\mathscr{L}$ since $\mathscr{L}$ is maximal. Having shown that $\mathscr{L}$ is a filter, it remains to show that $\mathscr{L}$ is an ultrafilter. Supposing the contrary, there would exist a subset E of X such that neither E nor $(X - E)$ belongs to $\mathscr{L}$. Hence there are sets A_1 and A_2 in $\mathscr{L}$ such that neither $A_1 \cap E$ nor $A_2 \cap (X - E)$

belongs to $\mathscr{P}$. If $A = A_1 \cap A_2$, then $A \in \mathscr{P}$ while neither $A \cap E$ nor $A \cap (X - E)$ belongs to $\mathscr{P}$, a contradiction.

(5.8) THEOREM. *A collection σ of subsets of a proximity space (X, δ) is a cluster if and only if there exists an ultrafilter $\mathscr{L}$ in X such that*

(5.9) $$\sigma = \{A \subset X : A \, \delta \, B \quad \text{for every} \quad B \in \mathscr{L}\}.$$

Moreover, given σ and $A_0 \in \sigma$, there exists an $\mathscr{L}$ satisfying (5.9) which contains A_0.

Proof. Let $\mathscr{L}$ be an ultrafilter in X and let σ be defined by (5.9). We shall first show that σ is a cluster.

(i) Suppose A and B belong to σ. For every subset C, either C or $(X - C)$ is in $\mathscr{L}$. This means that both A and B are near to either C or $(X - C)$. Hence for every subset C, either $A \, \delta \, C$ or $(X - C) \, \delta \, B$ which shows (by the Strong Axiom) that $A \, \delta \, B$.

(ii) Since $\mathscr{L} \subset \sigma$, 5.4 (ii) follows easily.

(iii) If neither A nor B belongs to σ, there exist A' and B' in $\mathscr{L}$ such that $A \, \bar{\delta} \, A'$ and $B \, \bar{\delta} \, B'$. Using (1.2), we obtain

$$A \, \bar{\delta} \, (A' \cap B') \quad \text{and} \quad B \, \bar{\delta} \, (A' \cap B').$$

Thus $(A \cup B) \, \bar{\delta} \, (A' \cap B')$. Since $(A' \cap B') \in \mathscr{L}$, it follows that $(A \cup B) \notin \sigma$, as required.

Conversely, let σ be a cluster and suppose $A_0 \in \sigma$. Taking $\mathscr{P} = \sigma$ in Lemma (5.7), we obtain an ultrafilter $\mathscr{L} \subset \sigma$ such that $A_0 \in \mathscr{L}$. If $\sigma' = \{A \subset X : A \, \delta \, B$ for every $B \in \mathscr{L}\}$, then $\sigma \subset \sigma'$. Thus by Lemma (5.6), $\sigma = \sigma'$, and (5.9) is satisfied.

(5.10) COROLLARY. *If $\mathscr{L}$ is an ultrafilter such that $\mathscr{L} \subset \sigma$, then σ is uniquely determined.*

(5.11) REMARKS. (i) If $\mathscr{L}$ and σ are as in Theorem (5.8), we say that '$\mathscr{L}$ generates σ' or 'σ is determined by $\mathscr{L}$'.

(ii) In Theorem (5.8), $\mathscr{L}$ need only be an ultrafilter base.

(iii) A given cluster σ may be determined by different ultra-filters, as the following example illustrates:

Let X be the real line with the usual topology and let δ be defined by $A \, \delta \, B$ iff $\bar{A} \cap \bar{B} \neq \varnothing$. Then δ is a compatible proximity by Theorem (2.11). Now consider the point cluster σ_0. Let $\mathscr{L}_1$ and $\mathscr{L}_2$ be ultrafilters containing the filter bases

$$\mathscr{F}_1 = \{(-a,0): a > 0\} \quad \text{and} \quad \mathscr{F}_2 = \{(0,a): a > 0\}$$

respectively. Then $\mathscr{L}_1$ and $\mathscr{L}_2$ both generate σ_0, but $\mathscr{L}_1 \neq \mathscr{L}_2$.

(5.12) LEMMA. *If a cluster σ in a proximity space (X, δ) is determined by an ultrafilter $\mathscr{L}$, then σ is a point cluster σ_x if and only if $\mathscr{L}$ converges to x.*

Proof. $\sigma = \sigma_x$ iff $\{x\} \in \sigma$ iff $x \, \delta \, A$ for every $A \in \mathscr{L}$ iff x is a cluster point of $\mathscr{L}$ iff $\mathscr{L}$ converges to x.

A topological space is compact if and only if every ultrafilter in the space converges to a point. The following analogue of this result follows directly from the above lemma.

(5.13) THEOREM. *A proximity space is compact if and only if every cluster in the space is a point cluster.*

If $A \cap B \neq \varnothing$, then there exists an ultrafilter $\mathscr{L}$ which contains both A and B. A similar result holds for clusters in proximity spaces:

(5.14) THEOREM. *If $A \, \delta \, B$, then there exists a cluster σ in (X, δ) such that A and B both belong to σ.*

Proof. Let $\mathscr{P} = \{C \subset X : C \, \delta \, B\}$. From Lemma (5.7), there exists an ultrafilter $\mathscr{L}$ such that $A \in \mathscr{L} \subset \mathscr{P}$. The cluster σ determined by $\mathscr{L}$ contains both A and B.

(5.15) REMARKS. (i) Theorems (5.13) and (5.14) together yield the result: two subsets of a compact proximity space are near if and only if their closures intersect. Therefore, a compact completely regular space has a unique compatible proximity (cf. Theorem (3.7)).

(ii) If $A \, \delta \, B$, then there may be different clusters σ and σ' such that A and B are members of both. For example, let X be the real line with the usual topology and define

$$A \, \delta \, B \quad \text{iff} \quad \bar{A} \cap \bar{B} \neq \varnothing.$$

If $A = [0, 1]$ and $B = (0, 1)$, then $A \, \delta \, B$, and both sets belong to the point clusters σ_0 and σ_1.

We now prove a result similar to: if $\mathscr{L}$ is an ultrafilter in Y and $X \subset Y$, then the trace of $\mathscr{L}$ on X is an ultrafilter in X iff $X \in \mathscr{L}$.

(5.16)   THEOREM. *Let σ be a cluster in a proximity space (Y, δ) and let $X \in \sigma$. Then there exists a unique cluster in (X, δ_X) contained in σ, namely $\sigma' = \{A \subset X : A \in \sigma\}$.*

Proof. By Theorem (5.8), σ is determined by an ultrafilter $\mathscr{L}$ containing X. Then $\mathscr{L}_X = \{L \cap X : L \in \mathscr{L}\}$, the trace of $\mathscr{L}$ on X, is an ultrafilter in X and so generates a cluster σ' in X. If $A \in \sigma'$, then $A \, \delta \, (L \cap X)$ for each $L \in \mathscr{L}$. This implies $A \, \delta \, L$ for each $L \in \mathscr{L}$, i.e. $A \in \sigma$. Thus $\sigma' \subset \sigma$, and clearly

$$\sigma' = \{A \subset X : A \in \sigma\}.$$

That σ' is the only cluster in X contained in σ is shown by a method similar to that used in Lemma (5.6).

Let f be a function from X into Y. The following results are well known:

(a) If $\mathscr{L}$ is an ultrafilter base in X, then $f(\mathscr{L}) = \{f(L) : L \in \mathscr{L}\}$ is an ultrafilter base in Y.

(b) If X and Y are topological spaces, $\mathscr{L}$ is an O-filter base in X and f is continuous, then $f(\mathscr{L})$ is an O-filter base in Y. (Recall that $\mathscr{L}$ is an O-filter in (X, τ) iff for every $G \in \tau$, either $G \in \mathscr{L}$ or $(X - G) \in \mathscr{L}$.)

We find the following analogue in proximity spaces:

(5.17)   THEOREM. *If f is a proximity mapping from (X, δ_1) to (Y, δ_2), then to each cluster σ_1 in X, there corresponds a cluster σ_2 in Y such that*

$$\sigma_2 = \{A \subset Y : A \, \delta_2 f(B) \quad \text{for each} \quad B \in \sigma_1\}.$$

Proof. σ_1 is determined by an ultrafilter $\mathscr{L}$ in X. Now $f(\mathscr{L})$ is an ultrafilter base in Y and generates a cluster σ_2 in Y. If $A \, \delta_2 f(B)$ for every $B \in \sigma_1$, then $A \, \delta_2 f(L)$ for every $L \in \mathscr{L}$, so that $A \in \sigma_2$. To prove the reverse inclusion, we first note that

$$f(\sigma_1) \subset \sigma_2.$$

This follows from the fact that if $B \in \sigma_1$, then $B \, \delta_1 L$ for every $L \in \mathscr{L}$, and f being a proximity mapping implies $f(B) \, \delta_2 f(L)$ for each $L \in \mathscr{L}$, i.e. $f(B) \in \sigma_2$. Thus if $A \in \sigma_2$, then $A \, \delta_2 f(B)$ for every $B \in \sigma_1$.

The following result corresponds to the well-known theorem: if $\mathscr{L}$ is an ultrafilter in X and $X \subset Y$, then $\mathscr{L}$ is an ultrafilter base in Y and thus generates an ultrafilter in Y.

(5.18) COROLLARY. *If X is a subspace of a proximity space (Y, δ), then every cluster σ' in X is a subclass of a unique cluster σ in Y, and*

$$\sigma = \{A \subset Y : A \, \delta \, B \quad \textit{for every} \quad B \in \sigma'\}.$$

We now present an axiomatic characterization of the family of all clusters in a *separated* proximity space.

(5.19) DEFINITION. *A* semi-ultrafilter $\mathscr{S}$ *in a set X is a collection of non-empty subsets of X satisfying the following conditions:*

(i) *If $(A \cup B) \in \mathscr{S}$, then $A \in \mathscr{S}$ or $B \in \mathscr{S}$.*

(ii) *If $A \in \mathscr{S}$ and $A \subset B$, then $B \in \mathscr{S}$.*

(iii) *If $A \cap S \neq \varnothing$ for every $S \in \mathscr{S}$, then $A \in \mathscr{S}$.*

(iv) *If $\{x\}$ and $\{y\}$ both belong to $\mathscr{S}$, then $x = y$.*

Although every ultrafilter is clearly a semi-ultrafilter, the converse is not true. Every cluster in a separated proximity space is a semi-ultrafilter, showing that a semi-ultrafilter is not necessarily a filter.

(5.20) THEOREM. *Let $\mathscr{C}$ denote the family of all clusters in a separated proximity space (X, δ). Then the following conditions are satisfied:*

(i) *Let A and B be subsets of X. If for every $E \subset X$ there is a $\sigma \in \mathscr{C}$ such that either A, $E \in \sigma$ or $(X - E)$, $B \in \sigma$, then there is a $\sigma' \in \mathscr{C}$ such that A, $B \in \sigma'$.*

(ii) *Consider any $\sigma \in \mathscr{C}$. If for each $B \in \sigma$ there is a $\sigma' \in \mathscr{C}$ such that A, $B \in \sigma'$, then $A \in \sigma$.*

(iii) *Each ultrafilter in X is contained in some member of $\mathscr{C}$.*

Proof. (i) is a consequence of the Strong Axiom and Theorem (5.14). Condition (ii) follows from the fact that any two sets in a cluster are near to one another, and (iii) is proved in Theorem (5.8).

(5.21) THEOREM. *If $\mathscr{C}$ is a family of semi-ultrafilters in X satisfying conditions 5.20(i), (ii) and (iii), then there exists a separated proximity δ on X such that $\mathscr{C}$ is the family of all clusters in (X, δ).*

Proof. If $\mathscr{C}$ is a family of semi-ultrafilters satisfying 5.20(i), (ii) and (iii), define δ by:

(5.22) $A \, \delta \, B$ iff there is a $\sigma \in \mathscr{C}$ such that $A, B \in \sigma$.

(Theorem (5.14) provides a motivation for this definition.)

That δ is a separated proximity on X will now be verified.

(1) Symmetry is obvious from (5.22).

(2) $(A \cup B) \, \delta \, C$ iff there exists a $\sigma \in \mathscr{C}$ such that $(A \cup B)$, $C \in \sigma$. This is equivalent to $A, C \in \sigma$ or $B, C \in \sigma$, i.e. $A \, \delta \, C$ or $B \, \delta \, C$.

(3) That $A \, \bar{\delta} \, \varnothing$ follows from the fact that $\varnothing$ does not belong to any semi-ultrafilter.

(4) If for every subset E of X either $A \, \bar{\delta} \, E$ or $(X - E) \, \bar{\delta} \, B$, then there exists a $\sigma \in \mathscr{C}$ such that $A, E \in \sigma$ or $(X - E), B \in \sigma$. Since $\mathscr{C}$ satisfies 5.20(i) by hypothesis, there exists a $\sigma' \in \mathscr{C}$ such that $A, B \in \sigma'$, i.e. $A \, \delta \, B$.

(5) $A \cap B \neq \varnothing$ implies A and B both belong to some ultrafilter $\mathscr{L}$. Hence by 5.20 (iii), there exists a $\sigma \in \mathscr{C}$ such that $\mathscr{L} \subset \sigma$; that is, both A and B belong to σ, implying that $A \, \delta \, B$.

(6) If $x \, \delta \, y$, then x and y both belong to some $\sigma \in \mathscr{C}$. Then by 5.19(iv), $x = y$.

The next step is to prove that each $\sigma \in \mathscr{C}$ is a cluster in (X, δ).

(*a*) If $A, B \in \sigma$ then by (5.22), $A \, \delta \, B$.

(*b*) If $A \, \delta \, B$ for every $B \in \sigma$, then for each $B \in \sigma$ there exists a $\sigma' \in \mathscr{C}$ such that $A, B \in \sigma'$. By 5.20(ii), $A \in \sigma$.

(*c*) By 5.19(i), $(A \cup B) \in \sigma$ implies $A \in \sigma$ or $B \in \sigma$.

It now remains to show that every cluster σ in (X, δ) belongs to $\mathscr{C}$. From Theorem (5.8), we know that σ is determined by some ultrafilter $\mathscr{L}$ in X. By 5.20(iii), there is a $\sigma' \in \mathscr{C}$ such that $\mathscr{L} \subset \sigma'$. But σ' is a cluster and, since $\mathscr{L}$ generates a unique cluster, $\sigma = \sigma'$. Therefore, $\sigma \in \mathscr{C}$.

(5.23) REMARKS. Theorems (5.20) and (5.21) together indicate that the concept of a cluster may be considered as a primitive concept in proximity theory. We may then define a proximity space to be a set X together with a family of semi-ultrafilters satisfying 5.20(i), (ii) and (iii). Each member of this family will be called a cluster. This leads to an alternate treatment of proximity spaces and enables one to prove the compactification of a proximity space (see Section 7) without appealing to the axiom of choice.

6. Duality in proximity spaces

In Section 3, we saw an alternate way of studying proximity spaces using δ-neighbourhoods. If δ is a proximity on X, then δ and $\ll$ are dual relations on the power set of X in the following sense:

(6.1) DEFINITION. *Two relations β and β^* on the power set of X are dual iff $A\beta^* B$ is equivalent to $A\beta(X-B)$. Clearly $\beta^{**} = \beta$, justifying the term dual.*

In this section, a study is made of this duality with special reference to clusters and their duals, 'ends'.

(6.2) DEFINITION. *Let $\mathscr{G}$ be a class of subsets of a proximity space (X,δ). Define:*

(a) $\mathscr{G}^0 = \{E \subset X: \text{there exists an } A \in \mathscr{G} \text{ such that } A \ll E\}$.
(b) $\mathscr{G}' = \{E \subset X: E\,\delta\,A \text{ for every } A \in \mathscr{G}\}$.
(c) $\mathscr{G}^* = \{E \subset X: (X-E) \notin \mathscr{G}\}$, *the dual of $\mathscr{G}$.*

(6.3) LEMMA. *$\mathscr{G}^0$ and $\mathscr{G}'$ are dual classes.*
Proof. $E \in \mathscr{G}^0$ iff there exists an $A \in \mathscr{G}$ such that $A \ll E$. This is equivalent to the existence of an $A \in \mathscr{G}$ such that $(X-E)\,\delta\,A$, i.e. $(X-E) \notin \mathscr{G}'$.

(6.4) DEFINITION. *An end $\mathscr{F}$ in a proximity space (X,δ) is a collection of subsets of X satisfying the following conditions:*

(i) $B, C \in \mathscr{F}$ *implies the existence of a non-empty subset $A \in \mathscr{F}$ such that $A \ll B$ and $A \ll C$.*
(ii) *If $A \ll B$, then either $(X-A) \in \mathscr{F}$ or $B \in \mathscr{F}$.*

(6.5) DEFINITION. *A round filter $\mathscr{F}$ is a filter with the additional property that for each $F_1 \in \mathscr{F}$, there exists an $F_2 \in \mathscr{F}$ such that $F_2 \ll F_1$.*

(6.6) REMARKS. Clearly if $\mathscr{F}$ is an end in X, then $\varnothing \notin \mathscr{F}$ and $X \in \mathscr{F}$. An example of an end, and also of a round filter, is the system $\mathscr{N}_x$ of δ-neighbourhoods of a point $x \in X$. If $\mathscr{F}$ is a maximal round filter converging to x, then $\mathscr{F} = \mathscr{N}_x$.

Some authors use the term *regular filter* rather than round filter. A collection of sets $\mathscr{F}$ from a proximity space is said to be a *centred δ-system* iff (a) A, $B \in \mathscr{F}$ implies $A \cap B \neq \varnothing$, and (b) $A \in \mathscr{F}$ implies the existence of a $B \in \mathscr{F}$ such that $B \ll A$. Alexandroff originally defined an end to be a maximal centred δ-system. Theorem (6.9) shows the equivalence of this definition to the one stated in (6.4), noting that a maximal round filter is the same as a maximal centred δ-system.

(6.7) THEOREM. *Every end is a maximal round filter.*

Proof. Suppose that $\mathscr{F}$ is an end, and begin by showing that $\mathscr{F}$ is a filter. Condition 6.4(i) and the fact that $X \in \mathscr{F}$ together show that $\mathscr{F}$ is a non-empty filter base. Given that $C \in \mathscr{F}$ and $C \subset D$, we must show that $D \in \mathscr{F}$. By 6.4(i), there is an $A \in \mathscr{F}$ such that $A \ll C$. From 3.9(iii) we have $A \ll D$, and condition 6.4(ii) demands that either $(X - A) \in \mathscr{F}$ or $D \in \mathscr{F}$. But 6.4(i) excludes the first possibility since $A \in \mathscr{F}$, so that $D \in \mathscr{F}$.

That $\mathscr{F}$ is a round filter follows immediately from 6.4(i).

Finally, we must show that the round filter $\mathscr{F}$ is maximal. Let $\mathscr{G}$ be a round filter such that $\mathscr{F} \subset \mathscr{G}$. Suppose $B \in \mathscr{G}$. Then by (6.5), there is an $A \in \mathscr{G}$ such that $A \ll B$. Since $\mathscr{G}$ is a filter and $A \in \mathscr{G}$, we know $(X - A) \notin \mathscr{F}$. Hence $B \in \mathscr{F}$ by 6.4(ii), showing that $\mathscr{F} = \mathscr{G}$.

(6.8) LEMMA. *Let $\mathscr{F}$ be a round filter and $A \ll B$. If A intersects every member of $\mathscr{F}$, then B belongs to some round filter finer than $\mathscr{F}$.*

Proof. Let $\mathscr{G} = \{A \cap F : F \in \mathscr{F}\}$ and consider $\mathscr{G}^0$. We shall show that $\mathscr{G}^0$ is a round filter finer than $\mathscr{F}$ and that it contains B. Let P and Q be members of $\mathscr{G}^0$. Then, by definition, there exist members C and D of $\mathscr{F}$ such that $(A \cap C) \ll P$ and $(A \cap D) \ll Q$. Since $\mathscr{F}$ is a filter, $E = C \cap D \in \mathscr{F}$. From 3.9(iii) and 3.9(iv), it is evident that $(A \cap E) \ll P$, $(A \cap E) \ll Q$, and hence $(A \cap E) \ll (P \cap Q)$. Since $(A \cap E) \in \mathscr{G}$, it follows that $(P \cap Q) \in \mathscr{G}^0$. Hence $\mathscr{G}^0$ is a filter, since supersets of members of $\mathscr{G}^0$ clearly belong to $\mathscr{G}^0$. Now by 3.9(vi), there is an R such that $(A \cap E) \ll R \ll (P \cap Q)$. By taking $P = Q$ and noting that $(A \cap E) \in \mathscr{G}$ implies $R \in \mathscr{G}^0$, we see that the filter $\mathscr{G}^0$ is round.

Since $A \ll B$, we have $A \cap E \ll B$ by 3.9(iii), so that $B \in \mathscr{G}^0$. Finally, to show that $\mathscr{G}^0$ is finer than $\mathscr{F}$, suppose $E \in \mathscr{F}$. Since $\mathscr{F}$ is a round filter, there exists $F \in \mathscr{F}$ such that $F \ll E$. By 3.9 (iii), $A \cap F \ll E$ and so $E \in \mathscr{G}^0$.

(6.9) THEOREM. $\mathscr{F}$ *is an end if and only if it is a maximal round filter.*

Proof. In view of Theorem (6.7), it is sufficient to show that every maximal round filter $\mathscr{F}$ is an end. Condition 6.4 (i) is clearly satisfied by any round filter. In verifying 6.4(ii), suppose $A \ll B$ and $B \notin \mathscr{F}$. Since $\mathscr{F}$ is maximal, Lemma (6.8) guarantees the existence of an $E \in \mathscr{F}$ such that $A \cap E = \varnothing$. Thus $E \subset (X - A)$ and $(X - A) \in \mathscr{F}$ since $\mathscr{F}$ is a filter, proving 6.4(ii).

(6.10) COROLLARY. *Every round filter is a subclass of some end.*

The following is the main result of this section, pointing out the duality between clusters and ends.

(6.11) THEOREM. $\mathscr{F}$ *is an end if and only if* $\mathscr{F}^*$ *is a cluster.*

Proof. If $\mathscr{F}^*$ is a cluster, then:

(a^*) $A, B \in \mathscr{F}^*$ implies $A \delta B$.

(b^*) $A \delta B$ for every $B \in \mathscr{F}^*$ implies $A \in \mathscr{F}^*$.

(c^*) $(A \cup B) \in \mathscr{F}^*$ implies $A \in \mathscr{F}^*$ or $B \in \mathscr{F}^*$.

The dual $\mathscr{F}$ of $\mathscr{F}^*$ satisfies the following conditions:

(a) $A \ll B$ implies $(X - A) \in \mathscr{F}$ or $B \in \mathscr{F}$.

(b) Given $B \in \mathscr{F}$, there is an E such that $(X - E) \notin \mathscr{F}$ and $E \ll B$.

(c) $A, B \in \mathscr{F}$ implies $A \cap B \in \mathscr{F}$.

Note that (a) is simply a restatement of 6.4(ii).

If $\mathscr{F}$ is an end, it clearly satisfies (a) and (c). That (b) is satisfied follows from 6.4(i), which states that given $B \in \mathscr{F}$, there exists an $E \in \mathscr{F}$ such that $E \ll B$. Since $\mathscr{F}$ is a round filter by (6.7), $(X - E) \notin \mathscr{F}$.

Conversely, we must show that if $\mathscr{F}$ satisfies (a), (b) and (c), then $\mathscr{F}$ is an end. In doing so, we need only verify 6.4(i). In view of (c) and 3.9(iii), it is sufficient to show that given a $B \in \mathscr{F}$, there exists a non-empty $A \in \mathscr{F}$ such that $A \ll B$. If $B \in \mathscr{F}$, then (b) assures the existence of an E such that $E \ll B$ and $(X - E) \notin \mathscr{F}$. By 3.9(vi), there exists an A such that $E \ll A \ll B$. Since $E \ll A$ and $(X - E) \notin \mathscr{F}$, we see from (a) that $A \in \mathscr{F}$. Finally, $(X - E) \notin \mathscr{F}$ implies $E \neq \varnothing$, and so $A \neq \varnothing$.

(6.12) COROLLARY. *For discrete proximity spaces, ends and clusters both coincide with ultrafilters.*

(6.13) COROLLARY. $\mathscr{G}^0$ *is an end if and only if* $\mathscr{G}'$ *is a cluster.*

(6.14) THEOREM. *Every ultrafilter* $\mathscr{F}$ *in a proximity space contains a unique end* $\mathscr{F}^0$.

Proof. By Corollary (5.10), $\mathscr{F}$ is a subclass of a unique cluster, which in this case is $\mathscr{F}'$. Taking duals, we obtain $\mathscr{F}'^* \subset \mathscr{F}^*$. Using Lemma (6.3) and the self-duality of $\mathscr{F}$, this reduces to the statement that $\mathscr{F}^0 \subset \mathscr{F}$. That $\mathscr{F}^0$ is an end follows from Corollary (6.13).

To prove uniqueness, suppose that $\mathscr{G}$ is any end contained in $\mathscr{F}$. Then by (6.11) $\mathscr{G}^*$ is a cluster containing $\mathscr{F}$, and hence equals $\mathscr{F}'$. Thus $\mathscr{G} = \mathscr{F}'^* = \mathscr{F}^0$, by (6.3).

The following is the dual of Theorem (5.14):

(6.15) THEOREM. $A \ll B$ *in a proximity space* (X, δ) *if and only if every end in X contains either* $(X - A)$ *or B.*

Proof. Necessity is supplied by (6.4)(ii). To prove the converse, suppose $A \not\ll B$, i.e. $A \, \delta \, (X - B)$. By Theorem (5.14), there exists a cluster $\mathscr{F}^*$ containing both A and $(X - B)$. Hence both $(X - A)$ and B are not in $\mathscr{F}$ which, by Theorem (6.11), is an end.

The axiomatic characterization of clusters (Theorems (5.20), (5.21)) has the following analogue regarding ends:

(6.16) THEOREM. *Let* Φ *be a family of filters (called Φ-filters) in X. Then Φ is the family of all ends for some separated proximity on X if and only if the following conditions are satisfied:*

(i) *If $\mathscr{F}$ is a Φ-filter, then* $\bigcap_{F \in \mathscr{F}} F$ *contains at most one point.*

(ii) *If $\mathscr{F} \in \Phi$ and $B \in \mathscr{F}$, then there is an $A \in \mathscr{F}$ such that every Φ-filter contains either* $(X - A)$ *or B.*

(iii) *If every Φ-filter contains at least one of the sets C or D, then there is a B such that every Φ-filter which does not contain C*

contains B, and every Φ-filter which does not contain D contains
$(X - B)$.

(iv) *Every ultrafilter in X contains some Φ-filter as a subclass.*

Proof: Let Φ be the family of all ends in X. By (6.7), every end is a filter. 6.16(i) is a consequence of 6.4(ii) and 3.9(vii). In view of Theorem (6.15), 6.16(ii) follows from 6.4(i) while 6.16(iii) is equivalent to 3.9(vi). Statement 6.16(iv) is simply Theorem (6.14).

To prove sufficiency, suppose Φ is a family of filters satisfying 6.16(i)–(iv). Motivated by Theorem (6.15), we define, $A \ll B$ iff every Φ-filter contains either $(X - A)$ or B. Verification that $\ll$ satisfies 3.9(i)–(vii) is left to the reader. That every Φ-filter is an end follows from the definition of $\ll$ and 6.16(ii). Finally, since every end is contained in some ultrafilter, (6.14) and 6.16(iv) together imply that every end in X is a Φ-filter.

7. Smirnov compactification

It should be mentioned at the outset that throughout this section, we shall work exclusively with separated *proximity spaces.*

In Theorem (3.7), it was shown that every compact Hausdorff space has a unique compatible separated proximity. The problem considered in this section is to determine whether or not every separated proximity space (X, δ) can be embedded in a compact proximity space Y such that the proximity δ is the subspace proximity induced by the unique proximity on Y. This can indeed be done; in fact, there is an intimate relationship between the proximities associated with a Tychonoff space and its compactifications. A one-to-one order-preserving mapping (i.e. an order-isomorphism) exists between the two, and consequently the study of proximities can be reduced to that of compactifications. We have already seen (Remarks (2.18)) that the real line has two distinct compatible proximities. These actually correspond to two compactifications of the real line, namely the Alexandroff one-point compactification and the two-point compactification.

Using clusters, we shall now construct the Smirnov compactification of a separated proximity space. Let (X, δ) be a separated proximity space and let $\mathscr{X}$ denote the set of all clusters in X. For $A \subset X$, let

$$\overline{\mathscr{A}} = \{\sigma \in \mathscr{X} : A \in \sigma\}.$$

For $x \in X$, let $f(x) = \sigma_x$ (the point cluster). Then it is easy to see that

(i) f is a one-to-one mapping, and

(ii) $f(A) \subset \overline{\mathscr{A}}$.

It is to obtain property (i) that we insist on the proximity δ being separated, for we are then assured that each point $x \in X$ is a member of one and only one cluster in X.

(7.1) DEFINITION. *For $\mathscr{P} \subset \mathscr{X}$, we say that a subset A of X absorbs $\mathscr{P}$ iff $A \in \sigma$ for every $\sigma \in \mathscr{P}$, i.e. $\mathscr{P} \subset \overline{\mathscr{A}}$.*

(7.2) LEMMA. *The binary relation δ^* on the power set of $\mathscr{X}$ defined by*

(7.3) *$\mathscr{P} \delta^* \mathscr{Q}$ iff A absorbs $\mathscr{P}$ and B absorbs $\mathscr{Q}$ implies $A \delta B$, is a separated proximity on $\mathscr{X}$.*

Proof. (i) Symmetry of δ^* follows from that of δ.

(ii) Suppose that $\mathscr{Q} \delta^* \mathscr{R}$, that D absorbs $(\mathscr{P} \cup \mathscr{Q})$, and that C absorbs $\mathscr{R}$. Then D absorbs $\mathscr{Q}$ and hence $D \delta C$. Thus $(\mathscr{P} \cup \mathscr{Q}) \delta^* \mathscr{R}$. Conversely, suppose that $(\mathscr{P} \cup \mathscr{Q}) \delta^* \mathscr{R}$ and that $\mathscr{P} \delta^* \mathscr{R}$. Let B absorb $\mathscr{Q}$ and C absorb $\mathscr{R}$. Then we must show that $B \delta C$. Since $\mathscr{P} \delta^* \mathscr{R}$, there are sets A and D absorbing $\mathscr{P}$ and $\mathscr{R}$ respectively such that $A \delta D$. By the Strong Axiom, there is an E such that $A \delta E$ and $(X - E) \delta D$. Because D absorbs $\mathscr{R}$ and $(X - E) \delta D$, $(X - E)$ belongs to no cluster in $\mathscr{R}$. Consequently, $(C - E)$ belongs to no cluster in $\mathscr{R}$. But $C = (C - E) \cup (C \cap E)$ absorbs $\mathscr{R}$, implying that $C \cap E$ absorbs $\mathscr{R}$. Now $(A \cup B)$ absorbs $(\mathscr{P} \cup \mathscr{Q})$, which shows that $(A \cup B) \delta (C \cap E)$. Since $A \delta E$, we also have $A \delta (C \cap E)$ and hence $B \delta (C \cap E)$, implying that $B \delta C$.

(iii) That $\mathscr{P} \delta^* \mathscr{Q}$ implies $\mathscr{P}$ and $\mathscr{Q}$ are non-empty follows directly from (1.3).

(iv) If $\mathscr{P} \delta^* \mathscr{Q}$, there are sets A and B absorbing $\mathscr{P}$ and $\mathscr{Q}$ respectively such that $A \delta B$. By the Strong Axiom, there is an E such that $A \delta E$ and $(X - E) \delta B$. Since $(X - E) \delta B$ and B absorbs $\mathscr{Q}$, $(X - E)$ belongs to no cluster in $\mathscr{Q}$. Thus E absorbs $\mathscr{Q}$. Now let

$\mathscr{R} = \bar{\mathscr{E}}$. Then $\mathscr{P}\delta^*\mathscr{R}$ since A absorbs $\mathscr{P}$, E absorbs $\mathscr{R}$ and $A\,\delta\,E$. Since E belongs to no cluster in $(\mathscr{X} - \mathscr{R})$, $(X - E)$ absorbs $(\mathscr{X} - \mathscr{R})$. Therefore $(X - E)\,\delta\,B$ implies $(\mathscr{X} - \mathscr{R})\,\delta^*\,\mathscr{2}$.

(v) Suppose that $\mathscr{P} \cap \mathscr{2} \neq \varnothing$ and that A and B absorb $\mathscr{P}$ and $\mathscr{2}$ respectively. Then both A and B absorb $\mathscr{P} \cap \mathscr{2}$, so that $A\,\delta\,B$ and hence $\mathscr{P}\delta^*\mathscr{2}$.

(vi) A absorbs $\{\sigma\}$ where $\sigma \in \mathscr{X}$ iff $A \in \sigma$. Hence $\sigma_1\delta^*\sigma_2$ iff every set in σ_1 is in σ_2 and vice versa iff σ_1 and σ_2 coincide. Thus δ^* is a separated proximity on $\mathscr{X}$.

NOTATION. Let τ^* be the topology induced on $\mathscr{X}$ by δ^*.

(7.4) LEMMA. (X, δ) *is proximally isomorphic to* $f(X)$ *with the subspace proximity induced by* δ^*, *and* $f(X)$ *is dense in* $\mathscr{X}$.

Proof. We first note that A absorbs $f(B)$ iff $B \subset \bar{A}$ ($\tau(\delta)$-closure of A). Hence if $\mathscr{2}$ is any subset of $\mathscr{X}$, then

$\mathscr{2}\,\delta^*f(A)$ iff C absorbs $\mathscr{2}$ and D absorbs $f(A)$ implies $C\,\delta\,D$

$\qquad$ iff C absorbs $\mathscr{2}$ and $A \subset \bar{D}$ implies $C\,\delta\,D$

$\qquad$ iff C absorbs $\mathscr{2}$ implies $C\,\delta\,A$.

Taking $\mathscr{2}$ to be the singleton $\{\sigma\}$, we obtain

$\{\sigma\}\,\delta^*f(A)$ iff $C \in \sigma$ implies $C\,\delta\,A$

$\qquad$ iff $A \in \sigma$.

That is, $\bar{\mathscr{B}}$ is the τ^*-closure of $f(B)$. Therefore, since X belongs to every cluster, $f(X)$ is dense in $\mathscr{X}$.

Now $f(A)\,\delta^*f(B)$ iff $C\,\delta\,D$ whenever C and D absorb $f(A)$ and $f(B)$ respectively iff $C\,\delta\,D$ whenever $A \subset \bar{C}$ and $B \subset \bar{D}$. But this last statement is equivalent to $A\,\delta\,B$, so that $f(A)\,\delta^*f(B)$ iff $A\,\delta\,B$. Thus X is proximally isomorphic to $f(X)$.

(7.5) LEMMA. $(\mathscr{X}, \tau^*)$ *is compact.*

Proof. By Theorem (5.13), it suffices to show that an arbitrary cluster σ in $\mathscr{X}$ is a point cluster. Since $f(X)$ is dense in $\mathscr{X}$, 5.5(v) implies that $f(X) \in \sigma$. From Theorem (5.16), there is a unique cluster σ' in $f(X)$ such that $\sigma' \subset \sigma$. But X is proximally isomorphic to $f(X)$, so that there is a cluster σ'' in X which corresponds to σ'.

To be specific, $\sigma' = \{f(A): A \in \sigma''\}$. Now from the proof of Lemma (7.4), we know that $\{\sigma''\} \delta^* f(A)$ iff $A \in \sigma''$. Applying Corollary (5.18), we obtain $\sigma'' \in \sigma' \subset \sigma$, showing that σ is indeed a point cluster.

An alternative proof of the next lemma, which states that the compactification $(\mathscr{X}, \delta^*)$ of (X, δ) is essentially unique, is given in Remarks (7.9).

(7.6) LEMMA. *Any δ-homeomorphism g of (X, δ) onto a dense subset of a compact proximity space (Y, δ_1) extends to a δ-homeomorphism $\bar{g}$ of $(\mathscr{X}, \delta^*)$ onto (Y, δ_1).*

Proof. Suppose (X, δ) is δ-homeomorphic to a dense subset of a compact proximity space (Y, δ_1). To each cluster σ in X, there corresponds a cluster σ' in Y by Corollary (5.18). Since Y is compact, we know from Theorem (5.13) that σ' is a point cluster. By Theorem (5.16), every point (and hence every cluster) in Y determines a unique cluster (via the δ-homeomorphism of the dense subspace) in X. Thus the clusters in X are exactly those determined by points in Y, so that there is a bijection $\bar{g}$ from $\mathscr{X}$ onto Y which is an extension of g.

It remains to show that $\bar{g}$ is a δ-homeomorphism. If $\mathscr{P}$ and $\mathscr{Q}$ are subsets of $\mathscr{X}$ such that $\mathscr{P} \delta^* \mathscr{Q}$, then $\overline{\mathscr{P}} \cap \overline{\mathscr{Q}} \neq \varnothing$. Hence there exists a $\sigma \in \mathscr{X}$ such that $\{\sigma\} \delta^* \mathscr{P}$ and $\{\sigma\} \delta^* \mathscr{Q}$. Let $y = \bar{g}(\sigma)$. With the help of (1.4) and 5.4 (iii), we obtain $\{y\} \delta_1 \bar{g}(\mathscr{P})$ and $\{y\} \delta_1 \bar{g}(\mathscr{Q})$, whence $\bar{g}(\mathscr{P}) \delta_1 \bar{g}(\mathscr{Q})$. Conversely, $\bar{g}(\mathscr{P}) \delta_1 \bar{g}(\mathscr{Q})$ implies the existence of a point $y \in \overline{\bar{g}(\mathscr{P})} \cap \overline{\bar{g}(\mathscr{Q})}$, since Y is compact. Let $\sigma = \bar{g}^{-1}(y)$. Now if $A \in \sigma$ and B absorbs $\mathscr{P}$, then $A \delta \bar{g}(\mathscr{P})$ and $\bar{g}(\mathscr{P}) \subset \bar{B}$ (considering X as a subset of Y). But this implies $A \delta B$, so that $\{\sigma\} \delta^* \mathscr{P}$. Similarly, $\{\sigma\} \delta^* \mathscr{Q}$, from which we conclude that $\mathscr{P} \delta^* \mathscr{Q}$.

Combining the above sequence of lemmas we obtain the main result of this section:

(7.7) THEOREM. *Every separated proximity space (X, δ) is a dense subspace of a unique (up to δ-homeomorphism) compact Hausdorff space $\mathscr{X}$. Since $\mathscr{X}$ has a unique compatible separated proximity, subsets A and B of X are near iff their closures in $\mathscr{X}$ intersect ($\mathscr{X}$ is called the Smirnov compactification of X).*

(7.8)   COROLLARY. *The topology of a separated proximity space is Tychonoff.*

(7.9)   REMARKS. In the statement of Theorem (7.7), we have identified X with $f(X)$ as is usually done.

The Smirnov compactification of (X, δ) can also be constructed by embedding (X, δ) in the proximity space (Y, δ^{**}), where Y is the family of all ends in X and δ^{**} is defined in either of the two following equivalent ways:

(i) For subsets π_1 and π_2 of Y, $\pi_1 \delta^{**} \pi_2$ iff for every finite collection of sets $\{B_i : i = 1, ..., n\}$ such that $B_i \gg A_i$ where $\bigcup_{i=1}^{n} A_i = X$, there exist $F \in \pi_1$ and $G \in \pi_2$ such that $B_k \in (F \cap G)$ for some $1 \leqslant k \leqslant n$.

(ii) Setting $O(A) = \{\alpha : A \in \alpha, \alpha \in Y\}$, define $\pi_1 \delta^{**} \pi_2$ iff there exist sets A and B such that $A \, \delta \, B$, $\pi_1 \subset O(A)$ and $\pi_2 \subset O(B)$.

Since 7.9(ii) coincides with definition (7.3), it follows from (7.5) that $(Y, \tau(\delta^{**}))$ is compact. Since the Smirnov compactification is essentially unique, the duality existing between clusters and ends is again brought to our attention.

The following theorem was proved by Smirnov [136] in 1952:

In order that two compactifications of the space X be distinct, it is necessary and sufficient that there exist a pair of closed subsets of the space X, the closures of which intersect in one compactification and do not intersect in the other compactification.

This result may be used to provide a simple proof of the essential uniqueness of the compactification of a separated proximity space. For, suppose that a proximity space (X, δ) has two distinct compactifications (Y_1, δ_1) and (Y_2, δ_2). Then there does not exist a δ-homeomorphism between Y_1 and Y_2 which is the identity on X. Hence, using Theorem (4.4), there is no homeomorphism from Y_1 onto Y_2 which is the identity on X. Applying the above theorem together with Theorem (3.7), we then see that there must exist a pair of closed subsets of X which are near in one compactification but not near in the other. But this contradicts the fact that both Y_1 and Y_2 are δ-extensions of the proximity space (X, δ).

(7.10) THEOREM. *Every proximity mapping g of (X, δ_1) onto (Y, δ_2) has a unique extension to a continuous mapping $\bar{g}$ which maps the compactification of X onto the compactification of Y.*

Proof. It follows from Theorem (5.17) that if σ_1 is a cluster in X, there corresponds a cluster σ_2 in Y such that

$$\sigma_2 = \{P \subset Y : P\,\delta_2\,g(C) \quad \text{for every} \quad C \in \sigma_1\}.$$

Let $\bar{g}(\sigma_1) = \sigma_2$. Then $\bar{g}$ is a mapping from $\mathcal{X}$ to $\mathcal{Y}$. Clearly $\bar{g}$ maps the point cluster σ_x to the point cluster $\sigma_{g(x)}$. In other words, $\bar{g}$ agrees with g on X (identifying X with $f(X)$, where $f(x) = \sigma_x$).

In order to show that $\bar{g}$ is a proximity mapping, and hence continuous, we must show that $\mathcal{P}\,\delta_1^*\,\mathcal{Q}$ implies $\bar{g}(\mathcal{P})\,\delta_2^*\,\bar{g}(\mathcal{Q})$, i.e. if A absorbs $\bar{g}(\mathcal{P})$ and B absorbs $\bar{g}(\mathcal{Q})$ then $A\,\delta_2\,B$. If $A\,\bar{\delta}_2 B$, then by Lemma (3.3) there are subsets C and D of Y such that

$$A\,\bar{\delta}_2\,(Y-C), \quad B\,\bar{\delta}_2\,(Y-D) \quad \text{and} \quad C\,\bar{\delta}_2 D.$$

Since A absorbs $\bar{g}(\mathcal{P})$, $(Y-C)$ belongs to no cluster in $\bar{g}(\mathcal{P})$. But g is a proximity mapping, so that $g^{-1}(Y-C) = (X - g^{-1}(C))$ belongs to no cluster in $\mathcal{P}$. This shows that $g^{-1}(C)$ absorbs $\mathcal{P}$. Similarly, $g^{-1}(D)$ absorbs $\mathcal{Q}$. Since $\mathcal{P}\,\delta_1^*\,\mathcal{Q}$, we must have

$$g^{-1}(C)\,\delta_1\,g^{-1}(D).$$

But g is a proximity mapping, yielding $C\,\delta_2\,D$, a contradiction. Therefore $\bar{g}$ must be a proximity mapping.

That $\bar{g}(\mathcal{X}) = \mathcal{Y}$ follows from the facts that $f(Y) \subset \bar{g}(\mathcal{X}) \subset \mathcal{Y}$, $f(Y)$ is dense in $\mathcal{Y}$, and $\bar{g}(\mathcal{X})$ is compact. Therefore $\bar{g}$ is a mapping onto $\mathcal{Y}$.

Finally, we show that $\bar{g}$ is unique. Suppose $\bar{g}' \neq \bar{g}$ is another continuous extension of g mapping $\mathcal{X}$ onto $\mathcal{Y}$. Then there is a $\sigma \in \mathcal{X}$ such that $\bar{g}(\sigma) \neq \bar{g}'(\sigma)$. Since $\mathcal{Y}$ is Hausdorff and $\bar{g}$ is continuous, there is a neighbourhood $\mathcal{E}$ of σ such that

$$\bar{g}(\mathcal{E}) \cap \bar{g}'(\mathcal{E}) = \varnothing.$$

Now $f(X)$ is dense in $\mathcal{X}$, implying that there exists a point cluster $\sigma_x \in \mathcal{E} \cap f(X)$. For such a $\sigma_x, \bar{g}(\sigma_x) \neq \bar{g}'(\sigma_x)$. Therefore $\bar{g}$ and $\bar{g}'$ do not agree on $f(X)$, and hence not on X.

As is well known, compactifications of a topological space X may be partially ordered in the following manner:

$X_1^* \geqslant X_2^*$ iff the identity mapping of X can be extended to a continuous mapping of X_1^* onto X_2^*.

We have already defined, in Section 2, a partial order on the set of all proximities on a set X. The following important result follows directly from Theorems (3.7) and (7.10).

(7.11) THEOREM. *Given any Tychonoff space X, the Smirnov compactification defines an order-isomorphism u of the partially ordered set $\{\delta_i : i \in I\}$ of compatible proximities onto the partially ordered set $\{X_i^* : i \in I\}$ of Smirnov compactifications. That is, $\delta_i > \delta_j$ iff $X_i^* \geqslant X_j^*$ where $u(\delta_k) = X_k^*$ for $k \in I$.*

The above theorem is actually valid for a completely regular space as well. Following is an analogue of Urysohn's lemma for normal spaces:

(7.12) THEOREM. *In a proximity space (X, δ), $A \not\mathbin{\delta} B$ implies that there exists a proximity mapping $g: X \to [0, 1]$ such that $g(A) = 0$ and $g(B) = 1$.*

Proof. If $A \not\mathbin{\delta} B$, then $\bar{A} \cap \bar{B} = \varnothing$ in $\mathscr{X}$. But $\mathscr{X}$ is normal so that by Urysohn's lemma, there exists a continuous function $\bar{g} : \mathscr{X} \to [0, 1]$ such that $\bar{g}(A) = 0$ and $\bar{g}(B) = 1$. By Theorem (4.4), $\bar{g}$ is a proximity mapping. Setting $g = \bar{g} | f(X)$ and identifying X with $f(X)$ where $f(x) = \sigma_x$, we obtain the required mapping.

The next result follows directly from Tietze's extension theorem and Theorem (7.10).

(7.13) THEOREM. *Let A be any subspace of a proximity space X and let g be a proximity mapping from A to $[0, 1]$. Then g can be extended to a proximity mapping $\bar{g} : X \to [0, 1]$.*

The compactification of a proximity space, which we have considered above, is a special case of the general theory of proximal extensions. We shall now consider briefly this general theory and show its relationship to the theory of compactifications.

(7.14) DEFINITION. *A proximal (or δ-) extension of a proximity space (X, δ) is a proximity space (Y, δ') such that $\bar{X} = Y$ and $\delta = \delta'_X$. A proximity space is maximal (or absolutely closed) iff it has no proper δ-extension.*

(7.15) THEOREM. *A separated proximity space (X, δ) is maximal if and only if every cluster in X is a point cluster.*

Proof. If X is not maximal, there is a proper δ-extension Y; let $\xi \in Y - X$. Then from Theorem (5.16), there exists a unique cluster σ in X which is a subclass of the point cluster σ_ξ in Y. Obviously σ is not a point cluster in X.

Conversely, if there are clusters in X which are not point clusters, then the proximal extension of X given by (7.4) is proper. Thus X is not maximal.

(7.16) COROLLARY. *A separated proximity space is maximal if and only if it is compact.*

Every Tychonoff space has a maximal compactification, namely the Stone–Čech compactification. The following theorem characterizes those separated proximity spaces which have minimal compactifications,

(7.17) THEOREM. *A Tychonoff space X has a minimal compactification if and only if it is locally compact.*

Proof. If X is locally compact, we may resort to the Alexandroff one-point compactification, which is the minimal compactification of X. (Recall that the one-point compactification is Hausdorff iff X is locally compact Hausdorff.)

Conversely, suppose that X has a minimal compactification X^*. In order to show that X^* is the Alexandroff one-point compactification, suppose there are two different points ξ and η in $(X^* - X)$. We can then construct a smaller compactification, X_1^*, than X^* by 'pasting' ξ and η together. Open sets in X_1^* are then those which do not contain ξ or η, and those which contain both ξ and η.

(7.18) COROLLARY. *Every Tychonoff space X has a maximal compatible proximity. It has a minimal compatible proximity if and only if it is locally compact.*

A separated proximity relation δ may be defined on a locally compact Hausdorff space by

(7.19) $A \,\delta\, B$ iff $\bar{A} \cap \bar{B} = \varnothing$ and either $\bar{A}$ or $\bar{B}$ is compact.

Indeed, (7.11) and (7.17) show that a space has a smallest compatible proximity only if it is locally compact Hausdorff, and in this case the proximity is given by (7.19). We have previously seen that the largest proximity which can be defined on a Tychonoff space is $A \, \delta \, B$ iff A and B are functionally distinguishable. If a Tychonoff space possesses a unique proximity, the maximal and minimal compactifications must clearly coincide. These considerations together with the preceding results yield the following important theorem:

(7.20) THEOREM. *A Tychonoff space X has a unique compatible proximity if and only if it is locally compact and every pair of non-compact closed subsets A and B of X are functionally indistinguishable.*

We now consider those topological spaces which have a unique compatible proximity given by

$$A \, \delta \, B \quad \text{iff} \quad \bar{A} \cap \bar{B} \neq \varnothing.$$

It has already been shown (Theorem (3.7)) that compact completely regular spaces are of this form.

(7.21) DEFINITION. *A separated proximity space (X, δ) is equinormal iff $\bar{A} \cap \bar{B} = \varnothing$ implies $A \, \delta \, B$.*

It follows from Theorem (7.12) that every equinormal proximity space is normal. The converse is not true, however, since in (2.18) we have $\bar{A} \cap \bar{B} = \varnothing$ although $A \, \delta_1 \, B$. The following is a generalization of Theorem (4.4).

(7.22) THEOREM. *A normal separated proximity space (X, δ) is equinormal if and only if every real-valued continuous function on X is a proximity mapping.*

Proof. The proof of necessity is exactly the same as that for Theorem (4.4).

To prove the converse, let X be a normal proximity space such that every real-valued continuous function on X is a proximity mapping. If $\bar{A} \cap \bar{B} = \varnothing$, then by Urysohn's lemma there exists a continuous function $f: X \to [0, 1]$ such that $f(\bar{A}) = 0$ and $f(\bar{B}) = 1$. By hypothesis, f is also a proximity mapping and so $\bar{A} \, \delta \, \bar{B}$, i.e. $A \, \delta \, B$.

(7.23) COROLLARY. *A normal separated proximity space (X, δ) is equinormal iff the proximity δ is induced by the Stone–Čech compactification of X.*

(7.24) REMARK. If $f\colon X \to Y$ is continuous, δ is the proximity on X corresponding to the Stone–Čech compactification $\mathscr{X}$, and δ' is any proximity on Y, then f is a proximity map with respect to δ and δ'.

8. Proximity weight and compactification

In the previous section we studied compactifications of proximity spaces and the various implications of this theory. In this section we shall define topological weight and proximity weight, and prove stronger results concerning compactifications. These developments will lead to some interesting metrization theorems concerning proximity spaces.

(8.1) DEFINITION. *The* topological weight $w(\tau)$ *of a topological space* (X, τ) *is the smallest cardinal number* $\alpha \geqslant \aleph_0$ *such that* τ *has a topological base* $\mathscr{B}$ *with* $|\mathscr{B}| \leqslant \alpha$.

It is easy to see that $w(\tau)$ is a topological invariant.

(8.2) DEFINITION. *A* proximity base $\mathscr{P}$ *for a proximity space* (X, δ) *is a subset of the power set of* X *such that* $A\,\delta\,B$ *implies there are members* U *and* V *of* $\mathscr{P}$ *such that* $A \subset U$, $B \subset V$ *and* $U\,\delta\,V$.

It is obvious from Lemma (2.8) that the collection of all closed sets, in a proximity space, forms a proximity base.

(8.3) LEMMA. *If* $\mathscr{P}$ *is a proximity base, then so is*

$$\mathscr{P}_0 = \{\mathrm{Int}\,U : U \in \mathscr{P}\}.$$

Proof. If $A\,\delta\,B$, then by (3.5) there are sets C and D such that $A \ll C$, $B \ll D$ and $C\,\delta\,D$. This in turn implies the existence of members U and V of $\mathscr{P}$ such that $C \subset U$, $D \subset V$ and $U\,\delta\,V$. But $A \subset \mathrm{Int}\,C \subset \mathrm{Int}\,U \subset U$ and $B \subset \mathrm{Int}\,D \subset \mathrm{Int}\,V \subset V$. Hence $\mathrm{Int}\,U\,\delta\,\mathrm{Int}\,V$, and the conclusion follows.

(8.4) COROLLARY. *Given a proximity space* (X, δ), $\tau(\delta)$ *is a proximity base.*

(8.5) LEMMA. *If* $\mathscr{P}$ *is a proximity base of open sets, then* $\mathscr{P}$ *is a topological base.*

Proof. If G is an open set and $x \in G$, then $x\,\delta\,(X-G)$. Hence there are sets U and V in $\mathscr{P}$ such that $x \in U$, $(X-G) \subset V$ and

$U \delta V$. Also $U \cap V = \varnothing$, which shows that $x \in U \subset (X - V) \subset G$. Thus $\mathscr{P}$ is a topological base.

(8.6) THEOREM. *Let $\mathscr{B}$ be a topological base for a compact proximity space (X, δ). Then $\mathscr{B}^* = \{B : B = \bigcup_{i \in I} B_i$, where $B_i \in \mathscr{B}$ and I is finite\} is a proximity base.*

Proof. From Lemma (2.8), $A \delta B$ implies $\bar{A} \cap \bar{B} = \varnothing$. Since X is regular, for each $x \in \bar{A}$ there is a $U_x \in \mathscr{B}$ such that

$$x \in U_x \subset \bar{U}_x \subset (X - \bar{B}).$$

Now $\bar{A}$ is compact, so that $\bar{A} \subset \bigcup_{i=1}^{r} U_{x_i} \subset \bigcup_{i=1}^{r} \bar{U}_{x_i} \subset (X - \bar{B})$. Using a similar argument, we obtain

$$\bar{B} \subset \bigcup_{j=1}^{s} V_{y_j} \subset \bigcup_{j=1}^{s} \bar{V}_{y_j} \subset \left(X - \bigcup_{i=1}^{r} \bar{U}_{x_i} \right).$$

Let $$U^* = \bigcup_{i=1}^{r} U_{x_i} \quad \text{and} \quad V^* = \bigcup_{j=1}^{s} V_{y_j}.$$

Then $U^* \in \mathscr{B}^*$, $V^* \in \mathscr{B}^*$, $A \subset U^*$, $B \subset V^*$ and $\bar{U}^* \cap \bar{V}^* = \varnothing$. Therefore $U^* \delta V^*$, and we conclude that $\mathscr{B}^*$ is a proximity base.

The following result is obvious.

(8.7) LEMMA. *Let $\mathscr{P}$ be a proximity base of X and let $Y \subset X$. Then $\mathscr{P}_Y = \{U \cap Y : U \in \mathscr{P}\}$ is a proximity base for the subspace Y.*

(8.8) LEMMA. *Let Y be a dense subspace of a proximity space X and let $\mathscr{Q}$ be a proximity base for Y. Then $\mathscr{P} = \{\bar{U} : U \in \mathscr{Q}\}$ is a proximity base for X.*

Proof. If A and B are subsets of X such that $A \delta B$, then by (3.5) or (8.4) there are open sets G and H such that $A \subset G$, $B \subset H$ and $G \delta H$. Hence $(G \cap Y) \delta_Y (H \cap Y)$, which implies there are members U and V of $\mathscr{Q}$ such that $(G \cap Y) \subset U \subset Y$, $(H \cap Y) \subset V \subset Y$ and $U \delta_Y V$. Since $\bar{Y} = X$, $G \subset \overline{(G \cap Y)}$ and $H \subset \overline{(H \cap Y)}$. Thus $G \subset \bar{U}, H \subset \bar{V}$ and $\bar{U} \delta \bar{V}$.

(8.9) DEFINITION. *The proximity weight $w(\delta)$ of a proximity space (X, δ) is the smallest cardinal number $\alpha \geqslant \aleph_0$ such that X has a proximity base $\mathscr{P}$ with $|\mathscr{P}| \leqslant \alpha$.*

Clearly $w(\delta)$ is a proximity invariant. Applying Corollary (8.4) and Lemma (8.5), we obtain the following result:

(8.10)   THEOREM. *In a proximity space* (X, δ) *with* $\tau = \tau(\delta)$,

$$w(\tau) \leqslant w(\delta) \leqslant 2^{w(\tau)}.$$

Using this theorem together with Theorem (8.6), we obtain:

(8.11)   COROLLARY. *For a compact proximity space* (X, δ), $w(\tau) = w(\delta)$.

(8.12)   REMARK. It should be pointed out that the converse of (8.11) is false. For an infinite discrete space, $w(\tau) = w(\delta)$ although the space is not compact.

The next lemma follows easily from Lemmas (8.7) and (8.8).

(8.13)   LEMMA. *If* (Y, δ) *is a subspace of* (X, δ^*), *then*

$$w(\delta) \leqslant w(\delta^*).$$

If Y is dense in X, their proximity weights are equal.

We shall now consider a strengthened form of Theorem (7.7): namely, that the compactification of a separated proximity space can be effected so as to preserve the proximity weight.

(8.14)   THEOREM. *Every separated proximity space* (X, δ) *with proximity weight θ can be embedded δ-homeomorphically as a dense subspace of a compact Hausdorff space X^* with the same proximity weight θ. X^* is unique up to δ-homeomorphism.*

This theorem follows from (7.7) and (8.13), but will be proved alternatively below in a sequence of lemmas. The construction used here reminds one of a similar one used in constructing the Stone–Čech compactification of a Tychonoff space.

Let $\mathscr{P}$ be a proximity base for X such that $|\mathscr{P}| \leqslant \theta$, and let $\mathscr{C} = \{(U, V): U, V \in \mathscr{P} \text{ and } U \,\delta\, V\}$. Then $|\mathscr{C}| \leqslant \theta$. We know from Theorem (7.12) that for each $\gamma = (U, V) \in \mathscr{C}$, there exists a proximity mapping $f_\gamma : X \to [0, 1]$ such that $f_\gamma(U) = 0$ and $f_\gamma(V) = 1$. Now let $K = \underset{\gamma \in \mathscr{C}}{\times} [0, 1]$ be the Tychonoff cube of $|\mathscr{C}|$-copies of $[0, 1]$. Then K is a compact Hausdorff space and so, by Theorem (3.7), has a unique proximity δ^* given by $A \,\delta^*\, B$ iff $\bar{A} \cap \bar{B} \neq \varnothing$. Define $f : X \to K$ by $f(x) = (f_\gamma(x))$.

4

(8.15) LEMMA. *The function f is one-to-one.*

Proof. If $x \neq y$, then $x \, \delta \, y$. Therefore there are members U and V of $\mathscr{C}$ such that $x \in U$, $y \in V$, $f_\gamma(U) = 0$ and $f_\gamma(V) = 1$, where $\gamma = (U, V)$. Hence $f(x) \neq f(y)$.

(8.16) LEMMA. $f^{-1}: f(X) \to X$ *is a proximity mapping*

Proof. If A and B are subsets of X such that $A \, \delta \, B$, then there are members U and V of $\mathscr{P}$ such that $A \subset U$, $B \subset V$ and $U \, \delta \, V$. Let $\gamma_0 = (U, V)$. Now $f(A) \subset \underset{\gamma \in \mathscr{C}}{\times} U^\gamma$ and $f(B) \subset \underset{\gamma \in \mathscr{C}}{\times} V^\gamma$, where $U^{\gamma_0} = \{0\}$, $V^{\gamma_0} = \{1\}$ and $U^\gamma = V^\gamma = [0, 1]$ for all other γ. The sets containing $f(A)$ and $f(B)$ are closed and disjoint, showing that $f(A) \, \delta^* f(B)$. Thus f^{-1} is a proximity mapping.

(8.17) LEMMA. *The function* $f: X \to f(X)$ *is a proximity mapping.*

Proof. Since $f_\gamma: X \to [0, 1]$ is a proximity mapping, Theorem (7.10) assures the existence of a unique extension $\bar{f}_\gamma: \mathscr{X} \to [0, 1]$ where $\bar{f}_\gamma$ is a proximity mapping on the Smirnov compactification $\mathscr{X}$ of X. Since $\bar{f}_\gamma$ is continuous for each $\gamma \in \mathscr{C}$, $\bar{f}: \mathscr{X} \to K$ is also continuous, where $\bar{f}(x) = (\bar{f}_\gamma(x))$. But $\mathscr{X}$ is compact, so that $\bar{f}$ is a proximity mapping. This, then, implies that f too is a proximity mapping, as $f = \bar{f}|X$.

(8.18) DEFINITION. *A proximity space is said to be* metrizable *iff it is δ-homeomorphic to a metric proximity space.*

(8.19) THEOREM. *A proximity space (X, δ) is proximally homeomorphic to a totally bounded metric space if and only if its proximity weight $w(\delta) = \aleph_0$.*

Proof. Suppose $w(\delta) = \aleph_0$. By Theorem (8.14), there exists a compactification $\mathscr{X}$ of X with proximity weight $\aleph_0$, which equals the topological weight $w(\tau)$. But a Tychonoff topological space with $w(\tau) = \aleph_0$ is metrizable. Now a compact metric space is totally bounded and total boundedness is hereditary, showing that X is a totally bounded metric space.

Conversely, if (X, δ) is proximally homeomorphic to a totally bounded metric space (Y, d) then Y^*, the completion of Y, is compact and hence separable. Now a separable metric space has a countable base. Therefore by (8.11) the topological weight, and hence the proximity weight, of both Y and Y^* is $\aleph_0$. Thus the proximity weight of X is $\aleph_0$ as well.

(8.20) COROLLARY. *Every proximity space with proximity weight $\aleph_0$ is metrizable.*

(8.21) THEOREM. *The proximity weight of a metric proximity space is $\aleph_0$ if and only if it is totally bounded.*

Proof. Sufficiency follows from Theorem (8.19). On the other hand, if a metric space (Y, d) has proximity weight $\aleph_0$ then Y is δ-homeomorphic to a totally bounded metric space (Y', d'). By (4.8) Y and Y' are uniformly isomorphic, so that Y too is totally bounded.

The above theorem is a particular case of a more general result concerning the proximity weight of a metric space.

(8.22) DEFINITION. *A subset E of a metric space (Y, d) is ϵ-discrete iff $d(x, y) \geqslant \epsilon$ for every pair of distinct points x, y of E.*

Given $\epsilon > 0$, there exists, by Zorn's lemma, a maximal ϵ-discrete subset E_ϵ of Y; that is, $E_\epsilon \subset E \subset Y$ and E being ϵ-discrete implies $E = E_\epsilon$.

(8.23) THEOREM. *Let (Y, d) be a metric space of topological weight ψ and proximity weight θ. For $n \in N$, let E_n be a maximal $1/n$-discrete subset of Y and let $|E_n| = \sigma_n$. Then*

$$\psi = \sum_{n=1}^{\infty} \sigma_n \quad and \quad \theta = \sum_{n=1}^{\infty} 2^{\sigma_n}.$$

Proof. The family
$$\left\{ S\left(x, \frac{1}{2n}\right) : x \in E_n \right\}$$

consists of pairwise disjoint non-empty open sets in Y. The cardinality of this family is σ_n, so that $\sigma_n \leqslant \psi$ for each $n \in N$. Hence

$$\sum_{n=1}^{\infty} \sigma_n \leqslant \aleph_0 \psi = \psi.$$

On the other hand,
$$\left\{ S\left(x, \frac{1}{n}\right) : x \in E_n, n \in N \right\}$$

covers Y and forms a topological base. The cardinality of this family is $\leqslant \sum_{n=1}^{\infty} \sigma_n$, so that $\psi \leqslant \sum_{n=1}^{\infty} \sigma_n$. Thus $\psi = \sum_{n=1}^{\infty} \sigma_n$.

Every pair of disjoint subsets of E_n are not near, so that every proximity base of E_n must contain all subsets of E_n. Thus the proximity weight of E_n is $\geqslant 2^{\sigma_n}$ for each $n \in N$, showing that $\theta \geqslant \sum\limits_{n=1}^{\infty} 2^{\sigma_n}$. On the other hand, if two subsets A and B of Y are not near, then $D(A, B) = \epsilon > 0$. Choose n such that $1/n < \epsilon/5$. Let

$$U = \bigcup_{x \in A} S\left(x, \frac{1}{n}\right) \quad \text{and} \quad V = \bigcup_{x \in B} S\left(x, \frac{1}{n}\right).$$

Then $A \subset U$, $B \subset V$ and $D(U, V) \geqslant 3\epsilon/5 > 0$, implying that $U \,\delta^* \, V$. It is therefore clear that a proximity base for Y can be constructed by forming, for all possible subsets of E_n and for each $n \in N$, unions of spheres

$$S\left(x, \frac{1}{n}\right)$$

with points of subsets of E_n as centres. This proximity base has cardinality $\leqslant \sum\limits_{n=1}^{\infty} 2^{\sigma_n}$, i.e. $\theta \leqslant \sum\limits_{n=1}^{\infty} 2^{\sigma_n}$. Thus $\theta = \sum\limits_{n=1}^{\infty} 2^{\sigma_n}$.

(8.24) COROLLARY. *A metric proximity space has proximity weight $\aleph_0$ or $\geqslant 2^{\aleph_0}$ according as it is totally bounded or not.*

The proofs of the next two theorems are omitted, as they involve techniques similar to those used above.

(8.25) THEOREM. *If $\sigma_n = \psi$ for at least one $n \in N$, then $\theta = 2^{\psi}$. This is the particular case when $\psi = \eta^{\aleph_0}$, where η is an arbitrary cardinal number.*

(8.26) THEOREM. *For a metric proximity space, $\theta = 2^{\psi}$ or ψ according as there exists or does not exist an $n \in N$ such that $\sigma_n = \psi$.* (The generalized continuum hypothesis is needed to prove this.)

9. Local proximity spaces

In Section 7 it was shown that a compact Hausdorff space $\mathscr{X}$ is (uniquely) determined by a dense proximity space (X, δ), and that $A \,\delta\, B$ in X iff $\bar{A} \cap \bar{B} \neq \varnothing$ where the closures are taken in $\mathscr{X}$. It will be shown in this section that a locally compact Hausdorff space Y is determined by a dense subspace X when we know not only the proximity of X, but also which sets in X have compact closures in Y. (Such sets may be called *bounded*.) Boundedness in topological spaces has been studied axiomatically by various authors since 1939. This section is devoted to a treatment of the interrelationships existing between proximities and bounded sets, and eventually leads to the concept of a local proximity space. It will be seen that such spaces can be embedded as dense subspaces of locally compact Hausdorff spaces.

First, however, we shall pursue a brief study of boundedness in topological spaces.

(9.1) DEFINITION. *A non-empty collection $\mathscr{B}$ of subsets of a topological space X is called a* boundedness *in X iff*

(i) $A \in \mathscr{B}$ *and* $B \subset A$ *implies* $B \in \mathscr{B}$, *and*

(ii) $\{A_i : i = 1, ..., n\} \subset \mathscr{B}$ *implies* $\bigcup\limits_{i=1}^{n} A_i \in \mathscr{B}$.

Elements of $\mathscr{B}$ are called bounded sets.

(9.2) REMARKS. (i) If (X, d) is a metric space, we may define a *metric boundedness*

$$\mathscr{B}_d = \{A \subset X : \sup_{x,\,y \in A} d(x, y) < \infty\}.$$

It is clear that a metric boundedness is a boundedness.

(ii) Given any topological space X, we may define a boundedness $\mathscr{B}$ on X in the following manner:

$$\mathscr{B} = \{A \subset X : A \text{ is finite}\}.$$

(iii) If $\mathscr{B}$ is any boundedness, then $\varnothing \in \mathscr{B}$.

(iv) The intersection of a non-empty collection of bounded sets is bounded.

(v) If X is bounded, then every subset of X is bounded.

(9.3) LEMMA. *If $\mathscr{B}$ is a boundedness in X and X is unbounded, then*
$$\mathscr{F} = \{F : (X - F) \in \mathscr{B}\}$$
is a filter.

Proof. Since X is unbounded, each $F \in \mathscr{F}$ is non-empty. If F_1 and F_2 both belong to $\mathscr{F}$, then $(X - F_1)$ and $(X - F_2)$ both belong to $\mathscr{B}$. Then by 9.1 (ii),

$$(X - F_1) \cup (X - F_2) = [X - (F_1 \cap F_2)] \in \mathscr{B};$$

in other words, $\qquad\qquad F_1 \cap F_2 \in \mathscr{F}$.

If $F \in \mathscr{F}$ and $F \subset G$, then $(X - G) \subset (X - F)$. Since $(X - F) \in \mathscr{B}$, 9.1 (i) demands that $G \in \mathscr{F}$. Thus $\mathscr{F}$ is a filter.

(9.4) COROLLARY. *If for every subset E of X either $E \in \mathscr{B}$ or $(X - E) \in \mathscr{B}$, then $\mathscr{F}$ is an ultrafilter.*

Throughout the sequel, we shall always suppose that $\mathscr{B}$ is a boundedness in a topological space X.

(9.5) DEFINITION. *A topological space is* locally bounded *iff each point of the space has a bounded neighbourhood.*

The next result is easily proved using standard techniques.

(9.6) THEOREM. *A compact subset of a locally bounded topological space is bounded.*

(9.7) DEFINITION. *A topological space with a boundedness $\mathscr{B}$ is* compactly bounded *iff*

$$\mathscr{B} = \{B \subset X : \bar{B} \text{ is compact}\}.$$

We then have the following obvious theorem:

(9.8) THEOREM. *A Hausdorff space X is locally compact if and only if X being compactly bounded implies X is locally bounded.*

(9.9) DEFINITION. *X is* boundedly compact (*'Montel' space*) *iff every closed bounded subset of X is compact.*

In proving the following theorem and corollary, we use two well-known characterizations of compactness: A subset E of a topological space X is compact iff every filter base in E has a cluster point in E iff every ultrafilter in E converges to a point in E.

(9.10) THEOREM. *X is boundedly compact if and only if every closed bounded filter base (i.e. a filter base consisting of closed bounded sets) has a cluster point.*

Proof. Suppose that X is boundedly compact and let $\mathscr{F}$ be any closed bounded filter base in X. Given any $F_0 \in \mathscr{F}$, consider the filter base $\mathscr{F}_0 = \{F \cap F_0 : F \in \mathscr{F}\}$. Since F_0 is compact, $\mathscr{F}_0$ has a cluster point $x_0 \in F_0$. Clearly x_0 is also a cluster point of $\mathscr{F}$, and necessity is proved.

Now let B be any closed bounded subset of X, and consider any filter base $\mathscr{F}$ in B. Then $\overline{\mathscr{F}} = \{\overline{F} : F \in \mathscr{F}\}$ is a closed bounded filter base in $B \subset X$ and, by hypothesis, must have a cluster point x_0. In fact $x_0 \in \overline{F}$ for each $F \in \mathscr{F}$, so that x_0 is also a cluster point of $\mathscr{F}$. Therefore every filter base in B has a cluster point, implying that B is compact. Thus X is boundedly compact.

(9.11) COROLLARY. *X is boundedly compact iff every ultrafilter containing a closed bounded filter base converges.*

(9.12) DEFINITION. *A local proximity space is a triple $(X, \alpha, \mathscr{B})$ where X is a set, $\mathscr{B}$ is a boundedness in X and α is a binary relation on the power set of X satisfying (with δ replaced by α) (1.1), (1.2), (1.3) and (1.5), in addition to the following axioms:*

(a) Let $A \subset X$ and $B \in \mathscr{B}$. If for every $C \in \mathscr{B}$ either $A \alpha C$ or $(X - C) \alpha B$, then $A \alpha B$.

(b) If $A \alpha B$, then there is a set $D \in \mathscr{B}$ such that $D \subset B$ and $A \alpha D$. If α also satisfies (1.6), it is said to be *separated*.

(9.13) REMARKS. (i) If X is bounded, then α is a proximity on X.

(ii) If (X, δ) is a proximity space and $\mathscr{B}$ denotes the power set of X, then $(X, \delta, \mathscr{B})$ is a local proximity space.

(iii) From the definition of a boundedness and 9.12(*b*), we deduce that every singleton set, and hence every finite subset of X, is bounded.

(iv) Let X denote the positive real numbers with the usual metric proximity δ, and let $\mathscr{F}$ be the filter generated by the filter base consisting of all right rays. Let $\mathscr{B}$ consist of all complements of members of $\mathscr{F}$ (i.e. sets which are bounded above), and define
$$A \alpha B \text{ iff } (E \cap A) \delta (E \cap B) \text{ for some } E \in \mathscr{B}.$$
It will be shown in Theorem (9.16) that, since $\mathscr{F}$ is a free round filter, $(X, \alpha, \mathscr{B})$ is a local proximity space.

(v) $B \in \mathscr{B}$ implies the existence of a $C \in \mathscr{B}$ such that

$$B \not\delta (X - C).$$

(vi) The following is equivalent to 9.12 (a), where $\ll$ is defined with respect to α.

If $A \in \mathscr{B}$ and $A \ll C$, there exists a $B \in \mathscr{B}$ such that

$$A \ll B \ll C.$$

(9.14) DEFINITION. *A filter $\mathscr{F}$ is free iff* $\bigcap\limits_{F \in \mathscr{F}} F = \varnothing$.

(9.15) DEFINITION. *Given a local proximity space $(X, \alpha, \mathscr{B})$, a binary relation β on the power set of X is said to* agree locally *with α iff β and α agree whenever either of the sets involved is bounded.*

(9.16) THEOREM. *Let $\mathscr{F}$ be a free round filter in a proximity space (X, δ). Define*
  (a) $\mathscr{B} = \{B \subset X : (X - B) \in \mathscr{F}\}$, *and*
  (b) $A \alpha B$ iff $(E \cap A) \delta (E \cap B)$ *for some $E \in \mathscr{B}$.*
Then $(X, \alpha, \mathscr{B})$ is a local proximity space and α agrees with δ locally.

Proof. That $\mathscr{B}$ is a boundedness is immediate from (a). To prove that α agrees with δ locally, suppose $A \delta B$ where $B \in \mathscr{B}$. Since $\mathscr{F}$ is a round filter, the duality between $\mathscr{F}$ and $\mathscr{B}$ yields an $E \in \mathscr{B}$ such that $B \ll E$ relative to δ. Now $B \delta [(X - E) \cap A]$, so that $B \delta (E \cap A)$. Since $(E \cap B) = B$, $(E \cap A) \delta (E \cap B)$. Then by definition, $A \alpha B$. That $A \alpha B$ implies $A \delta B$ is obvious.

The definition of α and the fact that $\{x\} \in \mathscr{B}$ for every $x \in X$ together imply that α satisfies (1.5). The remainder of the axioms for a local proximity space, with the possible exception of 9.12 (a), follow easily. We now prove the contrapositive of 9.12 (a), namely $B \in \mathscr{B}$ and $B \ll D$ implies there is a $C \in \mathscr{B}$ such that

$$B \ll C \ll D$$

relative to α. Suppose $B \in \mathscr{B}$ and $B \ll D$ relative to α. Then $B \ll D$ relative to δ also, for α and δ agree locally. Since δ is a proximity, there exists an A such that $B \ll A \ll D$ relative to δ. As before, choose an $E \in \mathscr{B}$ such that $B \ll E$. Setting $C = E \cap A$, we obtain the required result since $C \in \mathscr{B}$ and δ agrees locally with α.

The next result indicates how a proximity may be defined on a local proximity space. This is valuable since it enables one to derive properties of local proximity spaces from known properties of proximity spaces. The proof is trivial.

(9.17) THEOREM. *Let* $(X, \alpha, \mathscr{B})$ *be a local proximity space, and define* δ *by*

$$A \,\delta\, B \quad iff \quad A \,\alpha\, B, \quad or \quad A \notin \mathscr{B} \quad and \quad B \notin \mathscr{B}.$$

Then δ, *called the* Alexandroff extension *of* α, *is a proximity on* X.

The next corollary follows from the fact that, due to 9.13 (iii), $x \,\delta\, A$ iff $x \,\alpha\, A$.

(9.18) COROLLARY. (i) *If for* $A \subset X$, *we define* $\bar{A} = \{x : x \,\alpha\, A\}$, *then* X *becomes a completely regular topological space.*

(ii) $A \,\alpha\, B$ *iff* $\bar{A} \,\alpha\, \bar{B}$.

Observe that the proximity δ defined in the above theorem is the smallest binary relation on the power set of X which agrees locally with α.

(9.19) LEMMA. *Every local proximity space* $(X, \alpha, \mathscr{B})$ *is locally bounded.*

Proof. This lemma follows directly from 9.13 (iii) and the contrapositive of 9.12 (a), namely $B \in \mathscr{B}$ and $B \ll D$ implies there is a $C \in \mathscr{B}$ such that $B \ll C \ll D$ relative to α.

(9.20) LEMMA. *In a local proximity space* $(X, \alpha, \mathscr{B})$, *the closure of a bounded set is bounded.*

Proof. If $B \in \mathscr{B}$, then by 9.13 (vi) there exists a $C \in \mathscr{B}$ with $B \ll C \ll X$. Hence $\bar{B} \subset C$.

(9.21) LEMMA. *Every compact subset of a local proximity space is bounded.*

Proof. Since every local proximity space is locally bounded, we simply apply Theorem (9.6).

(9.22) LEMMA. *Let* $(X, \alpha, \mathscr{B})$ *be a local proximity space and define* δ *as in Theorem* (9.17). *If* X *is unbounded, then*

$$\sigma = \{A : A \notin \mathscr{B}\}$$

is a cluster in (X, δ).

Proof. By the definition of δ, any two sets in σ are near. From 9.13(v), it follows that if $A\,\delta\,C$ for every $C\in\sigma$, then $A\in\sigma$. If $A\notin\sigma$ and $B\notin\sigma$, then $A\in\mathscr{B}$ and $B\in\mathscr{B}$. Hence $(A\cup B)\in\mathscr{B}$, i.e. $(A\cup B)\notin\sigma$. Thus, σ is a cluster.

We now turn our attention to the *local compactification* of a local proximity space. This could be accomplished by considering the space of all *bounded clusters* (i.e. clusters determined by *bounded ultrafilters*, the latter being those ultrafilters containing at least one bounded set) and defining a suitable topology on this space. However, since a compactification of a proximity space has already been constructed in Section 7, and by Theorem (9.17) a proximity space is always associated with a local proximity space, this local compactification can be performed more easily.

(9.23) THEOREM. *Given a separated local proximity space $(X,\alpha,\mathscr{B})$, there exists a locally compact Hausdorff space L and a one-to-one map $f\colon X\to L$ satisfying the following conditions:*

(i) *$A\,\alpha\,B$ iff $\overline{f(A)}\cap\overline{f(B)} \neq \varnothing$ in L.*

(ii) *$B\in\mathscr{B}$ iff $\overline{f(B)}$ is compact in L.*

(iii) *$\overline{f(X)} = L$.*

Such a local compactification is unique; that is, given

$$f_i\colon X\to L_i \quad (i=1,2)$$

satisfying the above conditions, there exists a homeomorphism h from L_1 onto L_2 such that $f_2 = h\circ f_1$. The function f is onto iff X is boundedly compact. L is compact iff X is bounded. Conversely, if we are given an injective mapping $f\colon X\to L$ with L locally compact Hausdorff and α, $\mathscr{B}$ are defined by statements (i) and (ii) then $(X,\alpha,\mathscr{B})$ is a local proximity space.

Proof. Let δ be the Alexandroff extension of α (Theorem (9.17)). Let $(f,\mathscr{X})$ be the Smirnov compactification of (X,δ) (Theorem (7.8)); i.e. $\mathscr{X}$ is a compact Hausdorff space, $A\,\delta\,B$ iff $\overline{f(A)}\cap\overline{f(B)} \neq \varnothing$, and $\overline{f(X)} = \mathscr{X}$. If X is bounded, then $\delta = \alpha$ and we take $L = \mathscr{X}$. So we need only consider the case in which $X\notin\mathscr{B}$. Now $\mathscr{X}$ consists of all clusters in X and so, by Lemma (9.22), $\sigma = \{A : A\notin\mathscr{B}\}\in\mathscr{X}$. Clearly $\sigma = \{A : \sigma\in\overline{f(A)}\}$. In other words, $B \in \mathscr{B}$ iff $\overline{f(B)} \subset (\mathscr{X} - \{\sigma\})$. Let $L = \mathscr{X} - \{\sigma\}$. Then L is locally compact and Hausdorff. Since $\{x\} \in \mathscr{B}$ for each $x \in X$, $f(X) \subset L$. One can now readily see that 9.23 (ii) and (iii) follow. Moreover, α and δ are locally equivalent and, by 9.12(b), if $A \alpha B$ we may assume $A \in \mathscr{B}$ and $B \in \mathscr{B}$. As a result, we obtain 9.23 (i), which in turn implies that f is one-to-one. In order to complete the proof, we need the following result:

(9.24)   For every $q \in L$, there exists a $B \in \mathscr{B}$ such that $q \in \overline{f(B)}$.

This becomes apparent by noting that, since L is locally compact, q has a compact neighbourhood Q. Hence by 9.23 (ii), $B = f^{-1}(Q)$ is bounded and $q \in \overline{f(B)}$.

We now prove uniqueness. Let $\mathscr{X}_i = L_i \cup \{\sigma_i\}$ be the one-point compactification of L_i. Then $(f_i, \mathscr{X}_i)$ is the Smirnov compactification of (X, δ). By Theorem (7.8), the Smirnov compactification is unique and there is a homeomorphism h from $\mathscr{X}_1$ onto $\mathscr{X}_2$ such that $h \circ f_1 = f_2$. It suffices to show that $h(\sigma_1) = \sigma_2$. If B is a bounded set in X, $\overline{f_1(B)}$ is a compact subset of L_1 which does not contain σ_1. Since h is a homeomorphism, $h(\sigma_1) \notin h(\overline{f_1(B)}) = \overline{f_2(B)}$. Hence by (9.24), $h(\sigma_1) \notin L_2$, i.e. $h(\sigma_1) = \sigma_2$.

If f is onto and B is a closed bounded set in X, we must show that B is compact. By 9.23 (ii), it suffices to show that $f(B)$ is closed in $f(X)$. Now by 9.23 (i), $f(x) \in \overline{f(B)}$ iff $x \in \bar{B}$. Since B is closed, $f(x) \in \overline{f(B)}$ iff $x \in B$, i.e. $f(x) \in f(B)$. Hence $f(B)$ is closed.

Conversely, suppose that X is boundedly compact and that $q \in L$. Choose B as in (9.24). We may assume that B is closed or, in other words, compact. But f is continuous by 9.23 (i), so that $f(B)$ is also compact. Hence $q \in f(B)$, and f is onto.

The proof of the remainder of the theorem is routine.

We next prove an analogue of Theorem (7.10) on extensions of proximity mappings.

(9.25)   DEFINITION.   Let $(X_i, \alpha_i, \mathscr{B}_i)$ $(i = 1, 2)$ be two local proximity spaces and $f \colon X_1 \to X_2$ be a function. Then f is a local proximity mapping iff:

(i) $A \alpha_1 B$ implies $f(A) \alpha_2 f(B)$, and
(ii) $B \in \mathscr{B}_1$ implies $f(B) \in \mathscr{B}_2$.

It is easy to see that the composition of two local proximity mappings is a local proximity mapping. Moreover, a local proximity mapping carries bounded clusters into bounded clusters.

(9.26) THEOREM. *Let* $(X_i, \alpha_i, \mathscr{B}_i)$ $(i = 1, 2)$ *be two local proximity spaces with local compactifications* (f_i, L_i) $(i = 1, 2)$. *If* $f: X_1 \to X_2$ *is a local proximity mapping, then there exists a unique continuous mapping* $h: L_1 \to L_2$ *such that* $h \circ f_1 = f_2 \circ f$. *Conversely, if* $h: L_1 \to L_2$ *is continuous and* $h \circ f_1(X_1) \subset f_2(X_2)$, *then the mapping* $f = f_2^{-1} \circ h \circ f_1 : X_1 \to X_2$ *is a local proximity mapping.*

Proof. Let $g = f_2 \circ f$. Then $g: X_1 \to L_2$ satisfies

(9.27) $A \, \alpha_1 B$ implies $\overline{g(A)} \cap \overline{g(B)} \neq \varnothing$, and

(9.28) $B \in \mathscr{B}_1$ implies $\overline{g(B)}$ is compact.

Given $p \in L_1$, let $\mathscr{F} = \{\overline{gf_1^{-1}(M)} \subset L_2 : M$ a neighbourhood of $p\}$. Since $f_1(X_1)$ is dense in L_1, no member of $\mathscr{F}$ is empty. Moreover, the intersection of finitely many neighbourhoods of p is a neighbourhood of p. This together with the fact that some member of $\mathscr{F}$ is compact (by 9.23 (ii), (9.28) and the local compactness of L_1) shows that

(9.29) $\mathscr{F}$ has a non-void intersection.

To see that

(9.30) $q \in \overline{gf_1^{-1}(M)}$ for every neighbourhood M of p implies $p \in \overline{f_1 g^{-1}(N)}$ for every neighbourhood N of q,

suppose q has a neighbourhood N such that $p \notin \overline{f_1 g^{-1}(N)}$. Since L_1 is regular, there exists a neighbourhood M of p such that $\overline{M} \cap \overline{f_1 g^{-1}(N)} = \varnothing$, and we have $f_1^{-1}(M) \not\alpha_1 g^{-1}(N)$ in X_1, by 9.23 (i). Therefore $f_1^{-1}(M) \cap g^{-1}(N) = \varnothing$ and hence $gf_1^{-1}(M) \cap N = \varnothing$, so that $q \notin \overline{gf_1^{-1}(M)}$.

If s and t are distinct points in L_2, they have disjoint compact neighbourhoods S and T respectively. Now (9.27) implies that $g^{-1}(S) \not\alpha_1 g^{-1}(T)$, so that $\overline{f_1 g^{-1}(S)} \cap \overline{f_1 g^{-1}(T)} = \varnothing$. By (9.29) and (9.30), we then have that

(9.31) $\mathscr{F}$ intersects in a unique point q.

We are now in a position to define h. For $p \in L_1$, let $h(p) = q$, where q corresponds to p in the manner shown above.

Let N be any open neighbourhood of q. Now we know that $\mathscr{F}$, a class of closed sets directed by inclusion, has a compact member C and intersects in a unique point $q \in C$. Hence some member of $\mathscr{F}$ is disjoint from the compact set $C - N$, since $C - N \notin \mathscr{F}$. But C belongs to $\mathscr{F}$, so that there exists an $F \in \mathscr{F}$ such that $F \subset C \cap N \subset N$. That is, for some neighbourhood M of p,

$$\overline{gf_1^{-1}(M)} \subset N.$$

This, together with the fact that $h(M) \subset \overline{gf_1^{-1}(M)}$ for every open $M \subset L_1$ (by definition of h), implies the continuity of h.

If $p = f_1(x)$ for some $x \in X$, then $g(x) \in gf_1^{-1}(M)$ for every set M containing p. Thus $g(x) = hf_1(x)$, and we have $h \circ f_1 = g = f_2 \circ f$.

To show uniqueness, suppose $h' : L_1 \to L_2$ is a continuous mapping such that $h' \circ f_1 = f_2 \circ f$. For $p \in L_1$, let q be defined as above. Since L_2 is Hausdorff, $h'(p) = q$ will follow if we can show that $h'(p) \in N$ for every neighbourhood N of q. If N is any open neighbourhood of q, then from the second last paragraph, there exists an open neighbourhood M of p such that $\overline{h'f_1f_1^{-1}(M)} \subset N$. Since M is open in L_1 and $\overline{f_1(X_1)} = L_1$, $L_1 = \overline{M \cap f_1(X_1)} \cup (L_1 - M)$. But $f_1f_1^{-1}(M) = M \cap f_1(X_1)$, so that $M \subset \overline{f_1f_1^{-1}(M)}$. Thus by the continuity of h', $h'(p) \in h'(M) \subset h'(\overline{f_1f_1^{-1}(M)}) \subset \overline{h'f_1f_1^{-1}(M)} \subset N$.

Since f_2^{-1}, h and f_1 are local proximity mappings and the local proximity of mappings is preserved under composition, the converse follows easily.

Notes

5. A thorough treatment of ultrafilters may be found in Gaal [C], for example. Clusters were introduced by Leader [53]. Relationships between clusters and ultrafilters are given in Leader [54] and Mrówka [76], the latter presenting the axiomatic characterization of the family of all clusters in a proximity space. The treatment presented here is due to the authors [120]. A weak form of Theorem (5.8) is implicitly contained in the aforementioned paper of Mrówka.

6. This section owes its existence to the work of Leader [55]. An excellent survey of centred systems in topological spaces has recently been published by Iliadis and Fomin [H]. The centred system is a subbase for a filter, and is generally used by the Soviet school in place of the filter. The concept of an end was originated by Alexandroff [A], while both Freudenthal [B] and Alexandroff [A] defined a round filter. Smirnov [98]

then used these devices in his proximal extension theory. Alexandroff originally defined an end in a Tychonoff space X using such terms as 'centred system' and 'completely regular system'. In filter terminology, he would define an end to be a maximal completely regular filter, the latter being a maximal filter with an open base $\mathscr{B}$ such that for each $B \in \mathscr{B}$, there exists an $A \in \mathscr{B}$ with $A \subset B$ and a continuous function $f: X \to [0, 1]$ such that $f(\overline{A}) = 0$ and $f(X - B) = 1$. Clearly, this definition of an end coincides with that of Smirnov (see (6.4)) when the proximity in question is that defined by (1.11).

7. A general treatment of compactifications can be found in Kelley [J]. For an account of n-point compactifications, the reader is referred to Magill [K]. The results on proximal extensions are due to Smirnov [98], who was the first to expose the relationship between proximities and compactifications (which he effected using ends). The construction of the Smirnov compactification presented here is a modification of the method of Leader [53]. Equinormal proximity spaces were first investigated by Pervin [86]. For a discussion of uniformizable spaces with a unique uniformity (and hence a unique proximity), refer to Gál [D].

8. The material in this section was taken from Császár and Mrówka [13]. Taking advantage of the Smirnov compactification constructed in the previous section, a few proofs have been simplified. The topological theorem used in the proof of (8.17) can be found on page 91 of Kelley [J]. The definition of a proximity base given in this section is due to Császár and Mrówka [13]. It is somewhat unsatisfactory, however, in that a unique proximity is not necessarily associated with a given proximity base; for instance, the family of all closed subsets of X is a base for every compatible proximity on X. Njåstad [83] has eliminated this problem by defining a proximity base for (X, δ) to be a collection $\mathscr{B}$ of subsets of X satisfying:

(i) $A \, \delta \, B$ implies that there exist $C, D \in \mathscr{B}$ such that $A \subset C, B \subset D$ and $C \, \delta \, D$.

(ii) If A and B are disjoint members of $\mathscr{B}$, then $A \, \delta \, B$.

9. A comprehensive study of boundedness, on which the first part of this section is based, was carried out by Hu [F]. Taking proximity and boundedness as primitive concepts, Leader [59] defined and studied local proximity spaces. Recently, Leader [130] has considered the problem of finding general conditions under which a topological space is determined by a dense subspace possessing a local proximity structure, and has generalized Theorem (9.26).

CHAPTER 3

PROXIMITY AND UNIFORMITY

10. Proximity induced by a uniformity

In this chapter, we shall study questions concerning the relationship between uniform structures and proximity structures. Proximity structures lie between topological structures and uniform structures in the sense that all topological invariants are proximity invariants and all proximity invariants are uniform invariants; however, some uniform invariants, such as total boundedness and completeness, are not proximity invariants.

A uniform structure on X was first defined by Weil in terms of subsets of $X \times X$. Tukey later provided an alternate description of a uniform structure using covers of X. Although we shall adhere to the Weil approach here, it should be noted that Tukey's method is widely used by the Soviet school.

If $U \subset X \times X$ then $U^{-1} = \{(x, y) : (y, x) \in U\}$. Whenever $U = U^{-1}$ U is called *symmetric*. For subsets U, V of $X \times X$, $U \circ V = \{(x, z) :$ there exists a $y \in X$ such that $(x, y) \in V$ and $(y, z) \in U\}$. We inductively define $\overset{n}{V}$ for $n \in N$ by $\overset{1}{V} = V$, $\overset{n+1}{V} = \overset{n}{V} \circ V$. Let

$$\Delta = \{(x, x) : x \in X\}.$$

If $A \subset X$, then $U[A] = \{y : (x, y) \in U$ for some $x \in A\}$. For $x \in X$, $U[x] = U[\{x\}]$.

(10.1) DEFINITION. *A uniform structure (or uniformity)* $\mathscr{U}$ *on a set X is a collection of subsets (called* entourages) *of $X \times X$ satisfying the following conditions*:

(i) *Every entourage contains the diagonal* $\Delta = \{(x, x) : x \in X\}$.

(ii) *If $U \in \mathscr{U}$ and $V \in \mathscr{U}$, then $U \cap V \in \mathscr{U}$.*

(iii) *Given $U \in \mathscr{U}$, there exists a $V \in \mathscr{U}$ such that $V \circ V \subset U$* (triangle inequality).

(iv) *If $U \in \mathscr{U}$ and $U \subset V \subset X \times X$, then $V \in \mathscr{U}$.*

(v) *If $U \in \mathscr{U}$, then $U^{-1} \in \mathscr{U}$.*

The pair $(X, \mathscr{U})$ is called a uniform space.

A subfamily $\mathscr{B}$ of a uniformity $\mathscr{U}$ is a *base* for $\mathscr{U}$ iff each entourage in $\mathscr{U}$ contains a member of $\mathscr{B}$. A family $\mathscr{S}$ is a *subbase* for $\mathscr{U}$ iff the family of finite intersections of members of $\mathscr{S}$ is a base for $\mathscr{U}$. A uniformity $\mathscr{U}$ is *totally bounded* iff for each $U \in \mathscr{U}$, there exists a family of sets $\{A_i : 1 \leqslant i \leqslant n\}$ such that $X = \bigcup_{i=1}^{n} A_i$ and $A_i \times A_i \subset U$ for each $i = 1, \ldots, n$. If $\mathscr{U}, \mathscr{V}$ are two uniformities on X such that $\mathscr{U} \subset \mathscr{V}$, then we say that $\mathscr{V}$ is *finer* than $\mathscr{U}$ or that $\mathscr{U}$ is *coarser* than $\mathscr{V}$. The supremum of two uniformities $\mathscr{U}$ and $\mathscr{V}$, denoted by $\mathscr{U} \vee \mathscr{V}$, is that uniformity which has as a base $\{U \cap V : U \in \mathscr{U}, V \in \mathscr{V}\}$.

Just as we defined a (pseudo-) metric proximity in Section 1, it is possible to define a (pseudo-) metric uniformity. Given a (pseudo-) metric d on X, let $U_\epsilon = \{(x, y) \in X \times X : d(x, y) < \epsilon\}$. Then the collection $\{U_\epsilon : \epsilon > 0\}$ is a base for the (pseudo-) metric uniformity induced by d.

It can be shown that for each $x \in X$, $\{U[x] : U \in \mathscr{U}\}$ is a neighbourhood filter. Thus $\mathscr{U}$ generates a topology $\tau = \tau(\mathscr{U})$ on X. As is well known, this topology is always completely regular.

If $\mathscr{U}$ satisfies the additional condition

(vi) $\bigcap_{U \in \mathscr{U}} U = \Delta$,

then $\mathscr{U}$ is called a *Hausdorff* or *separated* uniformity. In this case, $\tau(\mathscr{U})$ is Tychonoff. Conversely, every (Tychonoff) completely regular space (X, τ) has a *compatible* (separated) uniformity, i.e. a uniformity $\mathscr{U}$ such that $\tau = \tau(\mathscr{U})$. Every uniformity has a base consisting of open (closed) symmetric members, and it is frequently more convenient to work with such a base for $\mathscr{U}$ rather than with $\mathscr{U}$ itself.

(10.2) **THEOREM.** *Every uniform space $(X, \mathscr{U})$ has an associated proximity $\delta = \delta(\mathscr{U})$ defined by*

(10.3) $A \delta B$ iff $(A \times B) \cap U \neq \varnothing$ for every $U \in \mathscr{U}$.

Furthermore, $\tau(\mathscr{U}) = \tau(\delta)$. If $\mathscr{U}$ is separated, then so is $\delta(\mathscr{U})$.

Proof. All the axioms for a proximity, except perhaps (1.4), are easily verified. To verify (1.4), suppose $A \bar{\delta} B$. Then there exists an entourage U such that $(A \times B) \cap U = \varnothing$. Now by

10.1(iii), there exists an entourage V such that $V \circ V \subset U$. Let $E = V^{-1}[B]$. Then $(A \times E) \cap V = \varnothing$ and

$$((X - E) \times B) \cap V = \varnothing, \quad \text{i.e.} \quad A \, \delta \, E \quad \text{and} \quad (X - E) \, \delta \, B.$$

To show that $\tau(\delta) = \tau(\mathcal{U})$, we observe that x is in the $\tau(\mathcal{U})$-closure of A iff $x \in U[A]$ for every entourage U iff $(x \times A) \cap U \neq \varnothing$ for every entourage U iff $x \, \delta \, A$, i.e. x is in the $\tau(\delta)$-closure of A. Finally, suppose that $\mathcal{U}$ is separated. If $x \, \delta \, y$, then $(x, y) \cap U \neq \varnothing$ for every entourage U. This implies $(x, y) \cap \Delta \neq \varnothing$, so that $x = y$. Thus δ is separated.

(10.4) REMARKS. (i) Instead of using (10.3), δ could equivalently be defined by

$$A \, \delta \, B \text{ iff } U[A] \cap U[B] \neq \varnothing \quad \text{for every} \quad U \in \mathcal{U}, \quad \text{or}$$

$$A \, \delta \, B \text{ iff } U[A] \cap B \neq \varnothing \quad \text{for every} \quad U \in \mathcal{U}.$$

Definition (10.3) has been used here and care has been taken in writing the above proof, because the proof in this form will essentially go through if we are dealing with quasi-uniformities and quasi-proximities.

(ii) Two different uniformities on X may induce the same proximity δ on X. For example, let X be the real line and $\mathcal{U}_1$ be the usual metric uniformity. Let $\mathcal{U}_2$ be the subspace uniformity on X induced by the (unique) uniformity of its Smirnov compactification corresponding to the usual metric proximity. Clearly, $\mathcal{U}_1$ and $\mathcal{U}_2$ induce the same (metric) proximity on X. However, since $\mathcal{U}_1$ is not totally bounded whereas $\mathcal{U}_2$ is totally bounded, $\mathcal{U}_1$ and $\mathcal{U}_2$ are different.

(iii) It is possible for different uniformities on X to induce the same topology and yet induce different proximities on X. This becomes immediately clear if one recalls the proximities defined in Remarks (2.18) and considers the subspace uniformities induced by the Smirnov compactifications of X corresponding to δ_1 and δ_2.

(iv) Since every (Tychonoff) completely regular space has a compatible (separated) uniformity, the above theorem provides an alternate proof of the fact that every (Tychonoff) completely regular space has a compatible (separated) proximity. The reader should note the incompatible use of the term 'compatible'! It is hoped that in each case the meaning will be clear from the context.

(v) It should be observed that the triangle inequality 10.1 (iii) corresponds to the Strong Axiom.

Given a subset A of X and an entourage U, $U[A]$ may be called a 'uniform neighbourhood' of A. The following result justifies the application of this term to a subset B when $A \ll B$:

(10.5)   THEOREM. *Let* $(X, \mathscr{U})$ *be a uniform space and let* $\delta = \delta(\mathscr{U})$. *Then* $A \ll B$ *if and only if there is an entourage* U *such that*

$$U[A] \subset B.$$

Proof. $A \ll B$ iff $A \,\delta\, (X - B)$ iff $(A \times (X - B)) \cap U = \varnothing$ for some $U \in \mathscr{U}$. But the last statement is equivalent to $U[A] \subset B$.

(10.6)   REMARK. The above result suggests an alternate definition of the associated proximity relation: $A \ll B$ iff $U[A] \subset B$ for some $U \in \mathscr{U}$.

(10.7)   THEOREM. *If* $\mathscr{U}_1 \subset \mathscr{U}_2$, *then* $\delta_1 < \delta_2$ *where* $\delta_i = \delta(\mathscr{U}_i)$ *for* $i = 1, 2$.

Proof. $A \,\delta_1\, B$ implies the existence of a $U_1 \in \mathscr{U}_1$ such that $(A \times B) \cap U_1 = \varnothing$. But $U_1 \in \mathscr{U}_2$, and thus $A \,\delta_2\, B$.

(10.8)   THEOREM. *If* $f : (X, \mathscr{U}_1) \to (Y, \mathscr{U}_2)$ *is uniformly continuous, then* $f : (X, \delta_1) \to (Y, \delta_2)$ *is proximally continuous where* $\delta_i = \delta(\mathscr{U}_i)$ *for* $i = 1, 2$.

Proof. Suppose on the contrary that $A \,\delta_1\, B$, but $f(A) \,\delta_2\, f(B)$. Then there exists a $U_2 \in \mathscr{U}_2$ such that $(f(A) \times f(B)) \cap U_2 = \varnothing$. Since f is uniformly continuous, there exists a $U_1 \in \mathscr{U}_1$ such that $(x, y) \in U_1$ implies $(f(x), f(y)) \in U_2$. But $A \,\delta_1\, B$, so that

$$(A \times B) \cap U_1 \neq \varnothing$$

which implies $(f(A) \times f(B)) \cap U_2 \neq \varnothing$, a contradiction.

(10.9)   REMARKS. The converse of the above theorem is not true. Consider the identity mapping $i : (X, \mathscr{U}_2) \to (X, \mathscr{U}_1)$ where

$X, \mathcal{U}_1$ and $\mathcal{U}_2$ are defined as in 10.4(ii). Then i is a proximity mapping from (X, δ) onto itself, but i is not uniformly continuous. However, as will be shown later, a proximity mapping is uniformly continuous under certain conditions: for example, when the uniformity of the domain space is pseudo-metrizable (12.20), or that of the range space is totally bounded (12.12). Moreover, given a proximity mapping $f: (X, \delta_1) \to (Y, \delta_2)$ and a uniformity $\mathcal{U}_2$ on Y such that $\delta_2 = \delta(\mathcal{U}_2)$, there exists a uniformity $\mathcal{U}_1$ on X such that $\delta_1 = \delta(\mathcal{U}_1)$ and $f: (X, \mathcal{U}_1) \to (Y, \mathcal{U}_2)$ is uniformly continuous.

11. Completion of a uniform space by Cauchy clusters

Let $(X, \mathcal{U})$ be a uniform space and $\delta = \delta(\mathcal{U})$. We have seen (Theorem (5.8)) that every ultrafilter in a proximity space generates a cluster and that given a set A in a cluster σ, there exists an ultrafilter containing A which generates σ. It therefore seems natural to call a cluster *Cauchy* if it is generated by a Cauchy ultrafilter (recall that a filter $\mathscr{F}$ is *Cauchy* iff for every $U \in \mathcal{U}$ there is an $F \in \mathscr{F}$ such that $F \times F \subset U$). One can readily convince oneself that every Cauchy cluster can be considered to be a *point cluster*, determined by a point of the completion of X (in which X is embedded). The neighbourhood system of this point will contain arbitrarily 'small' sets which intersect every member of the cluster. This leads to the following definition of a Cauchy cluster, which is equivalent to the above (as will be seen later), but is easier to work with. *Throughout this section, we shall suppose that $\mathcal{U}$ is an open symmetric base for a separated uniformity.*

(11.1) DEFINITION. *A cluster σ in $(X, \mathcal{U}, \delta)$ is* Cauchy *iff there exists a round Cauchy filter $\mathscr{M} \subset \sigma$ such that $M \cap C \neq \varnothing$ for each $M \in \mathscr{M}$ and $C \in \sigma$.*

(11.2) REMARKS. (i) Every point cluster in X is Cauchy. In fact, in the above definition we may take (as we shall always do in this section) $\mathscr{M}$ to be the neighbourhood filter of the point.

(ii) If a cluster is Cauchy, then every ultrafilter which generates it is Cauchy.

(iii) Given a round Cauchy filter $\mathcal{M}$, $U \in \mathcal{U}$ and $n \in N$, there exists an $M \in \mathcal{M}$ and $V \in \mathcal{U}$ such that $\overset{n}{V}[M] \times \overset{n}{V}[M] \subset U$ and $\overset{n}{V} \subset U$.

This can be seen from the following argument. Since $\mathcal{M}$ is Cauchy and round, there exist sets M', $M \in \mathcal{M}$ such that $M' \times M' \subset U$ and $M \ll M'$. Let $V_1 \in \mathcal{U}$ be such that $V_1[M] \subset M'$. Then V may be chosen to be that entourage satisfying $\overset{n}{V} \subset U \cap V_1$.

The following definition is easily seen to be equivalent to the usual one (see Remark (11.9)):

(11.3) DEFINITION. *A uniform space $(X, \mathcal{U})$ is complete iff every Cauchy cluster in $(X, \mathcal{U}, \delta)$ is a point cluster σ_x for some $x \in X$.*

(11.4) LEMMA. *A closed subspace Y of a complete uniform space $(X, \mathcal{U})$ is complete.*

Proof. The trace $\mathcal{U}_Y = \{U \cap (Y \times Y) : U \in \mathcal{U}\}$ of $\mathcal{U}$ on Y is a base for the subspace uniformity on Y. If σ_1 is any Cauchy cluster in Y, then a slight modification of the proof of Theorem (5.17) shows that σ_1 is a subclass of a unique Cauchy cluster σ_2 in X. Since X is complete, $\sigma_2 = \sigma_x$ for some $x \in X$. But $x \delta B$ for every $B \in \sigma_1$ and, since Y is closed, $x \in Y$. Therefore $\{x\} \in \sigma_1$.

Let f be a mapping which associates with each point $x \in X$, the point cluster σ_x. Then f is a one-to-one mapping of X onto the space $f(X)$ of all point clusters. Let X^* denote the set of all Cauchy clusters in X. From 11.2(i), it follows that

$$f(X) \subset X^* \subset \mathcal{X}.$$

For each Cauchy cluster σ, let $\mathcal{M}(\sigma)$ be one of the filters given by (11.1). For each $U \in \mathcal{U}$, define

$$U^* = \{(\sigma_1, \sigma_2) \in X^* \times X^* : \text{there exist} \quad M \in \mathcal{M}(\sigma_1),$$
$$N \in \mathcal{M}(\sigma_2) \quad \text{such that} \quad M \times N \subset U\}.$$

To see that U^* is independent of the choice of $\mathcal{M}(\sigma_1)$ and $\mathcal{M}(\sigma_2)$ suppose $(\sigma_1, \sigma_2) \in U^*$. Then by 11.2(iii), there exists an $M \in \mathcal{M}(\sigma_1)$, $N \in \mathcal{M}(\sigma_2)$ and a $V \in \mathcal{U}$ satisfying $V[M] \times V[N] \subset U$. Given any $\mathcal{M}'(\sigma_1) \neq \mathcal{M}(\sigma_1)$, we can find an $M' \in \mathcal{M}'(\sigma_1)$ such that $M' \times M' \subset V$. Now $M \cap M' \neq \varnothing$, so that $M' \subset V[M]$. Hence

$V[M] \in \mathscr{M}'(\sigma_1)$, $V[N] \in \mathscr{M}(\sigma_2)$ and $V[M] \times V[N] \subset U$, showing that U^* is well defined.

(11.5) LEMMA. $\mathscr{U}^* = \{U^* : U \in \mathscr{U}\}$ is a uniformity base on X^*.

Proof. Every U^* obviously contains the diagonal, and

$$(U \cap V)^* \subset U^* \cap V^*.$$

Given $U^* \in \mathscr{U}^*$, there exists a $V \in \mathscr{U}$ such that $V \circ V \subset U$. That $V^* \circ V^* \subset U^*$ follows from the following argument: if

$$(\sigma_1, \sigma_2) \in V^* \circ V^*,$$

then there exists a $\sigma_3 \in X^*$ such that $(\sigma_1, \sigma_3) \in V^*$ and $(\sigma_3, \sigma_2) \in V^*$. Hence there exists an $A \in \mathscr{M}(\sigma_1)$, $B \in \mathscr{M}(\sigma_2)$ and $C', C'' \in \mathscr{M}(\sigma_3)$ such that $A \times C' \subset V$ and $C'' \times B \subset V$. Setting $C = C' \cap C'' \in \mathscr{M}(\sigma_3)$, we have $A \times C \subset V$ and $C \times B \subset V$. Therefore

$$A \times B \subset V \circ V \subset U,$$

which implies $(\sigma_1, \sigma_2) \in U^*$.

Since $f(X) \subset X^* \subset \mathscr{X}$ and $f(X)$ is dense in $\mathscr{X}$ (by (7.4)), we have the following result:

(11.6) LEMMA. $f(X)$ is a dense subset of X^*.

Let δ^* be the proximity induced by $\mathscr{U}^*$ on X^*. The restriction $\mathscr{U}_f^*$ of $\mathscr{U}^*$ to $f(X)$ is a uniformity base on $f(X)$ and so induces the proximity δ_f^* on $f(X)$.

(11.7) LEMMA. $(X, \mathscr{U}, \delta)$ and $(f(X), \mathscr{U}_f^*, \delta_f^*)$ are proximally isomorphic.

Proof. Clearly f is one-to-one and onto. Suppose $A \delta B$. Given $U \in \mathscr{U}$, let $U_f^* = U^* \cap (f(X) \times f(X))$. Then we must show that $(f(A) \times f(B)) \cap U_f^* \neq \varnothing$. Let $V \in \mathscr{U}$ be such that $\overset{3}{V} \subset U$. Since $A \delta B$, there exist $a \in A$, $b \in B$ such that $(a, b) \in V$. Therefore $V[a] \times V[b] \subset U$, and the point clusters σ_a, σ_b satisfy the condition $(\sigma_a, \sigma_b) \in U_f^*$. Conversely, if $f(A) \delta_f^* f(B)$, then for each U_f^* (corresponding to an arbitrary $U \in \mathscr{U}$) there exists a

$$(\sigma_a, \sigma_b) \in U_f^*, \quad \text{where} \quad \sigma_a \in f(A), \ \sigma_b \in f(B).$$

Hence $(a, b) \in U$ and we have $(A \times B) \cap U \neq \varnothing$ for arbitrary $U \in \mathscr{U}$, showing that $A \delta B$.

It can similarly be proved that $(X, \mathscr{U})$ and $(f(X), \mathscr{U}_f^*)$ are uniformly isomorphic.

(11.8) LEMMA. *Every Cauchy cluster in $(X^*, \mathscr{U}^*, \delta^*)$ is a point cluster.*

Proof. Let σ^* be any Cauchy cluster in X^*. Since $f(X)$ is dense in X^*, a slight modification of the proof of Theorem (5.16) shows that σ^* determines a unique Cauchy cluster σ' in $f(X)$ such that $\sigma' \subset \sigma^*$. But σ' is isomorphic to a Cauchy cluster σ in X. In order to show that $\sigma \in \sigma^*$, it is sufficient to verify that for each $U^* \in \mathscr{U}^*$ and each $M \in \sigma'$, $(\sigma \times M) \cap U^* \neq \varnothing$. Given $U^* \in \mathscr{U}^*$, there exists a $V \in \mathscr{U}$ and $C \in \mathscr{M}(\sigma)$ such that $\overset{3}{V} \subset U$ and

$$C \times C \subset V.$$

Then $V[C] \times V[C] \subset U$. Setting $M_0 = V[C] \cap f^{-1}(M)$ we have $M_0 \in \sigma$ since $V[C] \in \mathscr{M}(\sigma)$, $f^{-1}(M) \in \sigma$, and we can find an ultra-filter containing both $V[C]$ and $f^{-1}(M)$ which generates σ. Choose a point $p \in M_0$. Since $V[C]$ is open, there exists a $W \in \mathscr{U}$ such that $W[p] \subset V[C]$. We therefore have $W[p] \times V[C] \subset U$, where $W[p] \in \mathscr{M}(\sigma_p)$, $V[C] \in \mathscr{M}(\sigma)$ and $\sigma_p \in M$. Thus $(\sigma_p, \sigma) \in U^*$ and $(\sigma \times M) \cap U^* \neq \varnothing$.

(11.9) REMARKS. The above result shows that $(X^*, \mathscr{U}^*, \delta^*)$ is complete; for if $\mathscr{F}$ is any Cauchy filter in X^*, then $\mathscr{F}$ is contained in a Cauchy ultrafilter. This ultrafilter generates a Cauchy cluster which, by Lemma (11.8), must be a point cluster σ_{x_0} for some $x_0 \in X^*$. Clearly x_0 is a cluster point of the Cauchy filter $\mathscr{F}$, and thus $\mathscr{F}$ converges to x_0.

Finally, we remark that every Cauchy cluster in X is generated by a Cauchy ultrafilter containing the neighbourhood filter of some point in X^*. (If the point is in $X^* - X$, consider the trace of its neighbourhood filter on X.) Hence Definition (11.1) is equivalent to: a cluster is Cauchy iff it is generated by a Cauchy ultrafilter.

12. Proximity class of uniformities

If (X, τ) is a Tychonoff space, then it is known that

(i) every compatible uniformity on X contains (as a subset) a compatible totally bounded uniformity;

(ii) there exists a largest compatible uniformity (called the *universal uniformity*);

(iii) it is locally compact iff there exists a smallest compatible uniformity on X.

In this section, we consider similar problems concerning a proximity space. Specifically, for a proximity space (X, δ), let $\Pi(\delta)$ (called the *proximity class of uniformities*) denote the set of all uniformities $\mathscr{U}$ on X such that $\delta = \delta(\mathscr{U})$. We are interested in the problems:

(a) Does $\Pi(\delta)$ have a smallest member?

(b) Does $\Pi(\delta)$ have a largest member?

We shall see that $\Pi(\delta)$ always has a smallest member, which is in fact the unique totally bounded member of $\Pi(\delta)$. It will also be shown, by means of an example, that $\Pi(\delta)$ need not have a largest member. For the special case in which δ is a pseudo-metric proximity, however, the corresponding pseudo-metric uniformity is the largest member of $\Pi(\delta)$. Necessary and sufficient conditions for $\Pi(\delta)$ to have a largest member will be discussed in the next section.

The following result is obvious and we omit the proof.

(12.1) LEMMA. *If $\mathscr{W}$ is a uniformity on X and $Y \subset X$, then $\delta_Y(\mathscr{W}) = \delta(\mathscr{W}_Y)$.*

(12.2) THEOREM. *Given a separated proximity space (X, δ), there exists a totally bounded member of $\Pi(\delta)$.*

Proof. By identification of δ-homeomorphic spaces, we may consider X to be a dense subspace of its Smirnov compactification $\mathscr{X}$. $\mathscr{X}$, being a compact Hausdorff space, has a unique compatible separated uniformity $\mathscr{W}^*$. From (12.1) and the fact that a compact uniform space is totally bounded (a hereditary property), it follows that the subspace uniformity $\mathscr{W} = \mathscr{W}_X^*$ is a member of $\Pi(\delta)$ and is totally bounded.

We now give an explicit construction for $\mathscr{W}$:

(12.3)   THEOREM. *Let (X, δ) be a separated proximity space and let $\mathscr{W}$ be any totally bounded uniformity belonging to $\Pi(\delta)$. Then the family $\mathscr{V}$ of sets of the form*

$$V = \cup \{A_k \times A_k : k = 1, \dots, n\},$$

where $A_k \gg B_k$ and $\bigcup_{k=1}^{n} B_k = X$, constitutes a uniformity base for $\mathscr{W}$.

Proof. Recall that by Theorem (10.5), $A_k \gg B_k (k = 1, \dots, n)$ iff there exist $W_k \in \mathscr{W}$ such that $W_k[B_k] \subset A_k (k = 1, \dots, n)$. Defining $W = \bigcap_{k=1}^{n} W_k \in \mathscr{W}$, we have $W[B_k] \subset A_k (k = 1, \dots, n)$. Hence for every $V \in \mathscr{V}$ there is a $W \in \mathscr{W}$ such that $W \subset V$, so that $\mathscr{V} \subset \mathscr{W}$. Conversely, given $W \in \mathscr{W}$, let $W_1 = W_1^{-1} \in \mathscr{W}$ satisfy $\overset{3}{W_1} \subset W$. Since $\mathscr{W}$ is totally bounded, there exists a family of sets

$$\{C_i : i = 1, \dots, m\} \quad \text{such that} \quad \bigcup_{i=1}^{m} C_i = X$$

and $C_i \times C_i \subset W_1 (i = 1, \dots, m)$. Clearly,

$$W_1[C_i] \times W_1[C_i] \subset \overset{3}{W_1} \subset W$$

for each i, so that $V_1 = \bigcup_{i=1}^{m} W_1[C_i] \times W_1[C_i] \in \mathscr{V}$ and is a subset of W. Thus $\mathscr{V}$ is a uniformity base for $\mathscr{W}$.

(12.4)   COROLLARY. *$\mathscr{W}$ is the smallest member of $\Pi(\delta)$.*

Proof. Let $\mathscr{U} \in \Pi(\delta)$ and $W \in \mathscr{W}$. By the above theorem, there exists a $V = \bigcup_{k=1}^{n} A_k \times A_k \subset W$, where $A_k \gg B_k$ and $\bigcup_{k=1}^{n} B_k = X$. Since $\mathscr{U} \in \Pi(\delta)$, there exists a $U \in \mathscr{U}$ such that

$$U[B_k] \subset A_k (k = 1, \dots, n).$$

Clearly, $U \subset \bigcup_{k=1}^{n} U[B_k] \times U[B_k] \subset V \subset W$, i.e. $W \in \mathscr{U}$.

(12.5)   COROLLARY. *$\mathscr{W}$ is the unique totally bounded member of $\Pi(\delta)$.*

(12.6) REMARKS. The family $\{A_k : k = 1, ..., n\}$ described in the preceding theorem, i.e. such that $A_k \gg B_k (k = 1, ..., n)$ where $\bigcup_{k=1}^{n} B_k = X$, is called a δ-*cover* of X.

In the above discussion we only considered a *separated* proximity. This restriction is, however, not necessary and it is possible to prove directly that $\mathscr{V}$ as defined in the statement of Theorem (12.3) is a base for a unique totally bounded uniformity which is compatible with a given proximity. It was in this way, in fact, that the result was first proved in the literature. However, in this approach the triangle inequality presents a little difficulty; our use of the Smirnov compactification avoids the manipulations.

Throughout the remainder of this section, $\mathscr{W} = \mathscr{W}(\delta)$ denotes the unique totally bounded member of $\Pi(\delta)$.

(12.7) THEOREM. *Let δ_i $(i = 1, 2)$ be two proximities on X and let $\mathscr{W}_i = \mathscr{W}(\delta_i)$. If $\delta_1 < \delta_2$, then $\mathscr{W}_1 \subset \mathscr{W}_2$.*

Proof. Given $W_1 \in \mathscr{W}_1$, there exists a symmetric member W_1' of $\mathscr{W}_1$ such that $\overset{3}{W_1'} \subset W_1$. Since $\mathscr{W}_1$ is totally bounded, there exists a family of sets $\{B_k : k = 1, ..., n\}$ such that $X = \bigcup_{k=1}^{n} B_k$ and $B_k \times B_k \subset W_1'$ for each k. Set $A_k = W_1'[B_k]$. Then $B_k \ll_1 A_k$ and, since $\delta_1 < \delta_2$, $B_k \ll_2 A_k$. Consequently, $W_2 = \cup A_k \times A_k \in \mathscr{W}_2$, by (12.3). Moreover, $W_2 \subset \overset{3}{W_1'} \subset W_1$, showing that $W_1 \in \mathscr{W}_2$.

(12.8) LEMMA. *Let (X, δ) be a proximity space and let $\mathscr{W} = \mathscr{W}(\delta)$. If V is a subset of $X \times X$ such that*

(12.9) $A \ll V[A]$ *for all* $A \subset X$,

then we also have

(12.10) $A \ll (V \cap W)[A]$ *for all* $W \in \mathscr{W}, A \subset X$.

Consequently, if $\delta^* < \delta$ and $\mathscr{U} \in \Pi(\delta^*)$, then $\mathscr{U} \vee \mathscr{W} \in \Pi(\delta)$.

Proof. In view of Theorem (12.3), we may assume that

$W = \bigcup\limits_{i=1}^{n} A_i \times A_i$, where $\{A_i : i = 1, \ldots, n\}$ is a δ-cover of X. We shall first verify (12.10) for $n = 2$, i.e. for

$$W = (A_1 \times A_1) \cup (A_2 \times A_2).$$

Since $\quad A = (A - A_2) \cup (A - A_1) \cup (A \cap A_1 \cap A_2),$

$$(V \cap W)[A] = (V \cap W)[A - A_2] \cup (V \cap W)[A - A_1]$$
$$\cup (V \cap W)[A \cap A_1 \cap A_2]$$
$$= (V[A - A_2] \cap A_1) \cup (V[A - A_1] \cap A_2)$$
$$\cup V[A \cap A_1 \cap A_2].$$

Using the distributive property, we obtain

$$(V \cap W)[A] = V[A] \cap (A_1 \cup V[A \cap A_2]) \cap (A_2 \cup V[A \cap A_1]).$$

Now $A - A_2 \subset X - A_2 \ll W[X - A_2] \subset A_1$ and, by hypothesis, $A \cap A_2 \ll V[A \cap A_2]$. With the help of (3.9) and (3.10), we obtain $A = (A - A_2) \cup (A \cap A_2) \ll A_1 \cup V[A \cap A_2]$ and similarly,

$$A \ll A_2 \cup V[A \cap A_1].$$

We also know that $A \ll V[A]$, so that $A \ll (V \cap W)[A]$.

For $n > 2$, we observe that $W = \cap W(I, J)$ where

$$W(I, J) = (A_I \times A_I) \cup (A_J \times A_J)$$

corresponding to a partition

$$\{I, J\} \quad \text{of} \quad \{1, \ldots, n\}, \quad A_I = \bigcup_{i \in I} A_i \quad \text{and} \quad A_J = \bigcup_{j \in J} A_j,$$

with the intersection being taken over all partitions. The result then follows by induction on n.

(12.11) THEOREM. *Let f be a proximity mapping from (X, δ_1) to (Y, δ_2). Given an arbitrary member $\mathscr{U}_2 \in \Pi(\delta_2)$, there exists a uniformity $\mathscr{U}_1 \in \Pi(\delta_1)$ such that $f : (X, \mathscr{U}_1) \to (Y, \mathscr{U}_2)$ is uniformly continuous. If $\mathscr{U}_2$ is totally bounded, then $\mathscr{U}_1$ may be chosen to be $\mathscr{W}(\delta_1)$.*

Proof. $(f \times f)^{-1}[\mathscr{U}_2]$ is a base for a uniformity $\mathscr{U}_1^*$ on X. It is known that f is uniformly continuous with respect to $\mathscr{U}_1^*$ and $\mathscr{U}_2$, and if $\mathscr{U}_2$ is totally bounded then so is $\mathscr{U}_1^*$. Let $\delta_1^* = \delta(\mathscr{U}_1^*)$. If

$A \delta_1^* B$, there exists a $U_1^* (= (f \times f)^{-1}[U_2]$, where $U_2 \in \mathscr{U}_2)$ belonging to $\mathscr{U}_1^*$ such that $(A \times B) \cap U_1^* = \varnothing$, i.e. $(f(A) \times f(B)) \cap U_2 = \varnothing$. This shows that $f(A) \delta_2 f(B)$. But $f: (X, \delta_1) \to (Y, \delta_2)$ is a proximity mapping, so that $A \delta_1 B$. Thus $\delta_1^* < \delta_1$. Set $\mathscr{U}_1 = \mathscr{U}_1^* \vee \mathscr{W}(\delta_1)$. Then by Lemma (12.8), $\mathscr{U}_1 \in \Pi(\delta_1)$. Clearly, the mapping

$$f: (X, \mathscr{U}_1) \to (Y, \mathscr{U}_2)$$

is uniformly continuous.

If $\mathscr{U}_2$ is totally bounded, then so is $\mathscr{U}_1^*$. Thus by (12.5),

$$\mathscr{U}_1^* = \mathscr{W}(\delta_1), \quad \text{implying that} \quad \mathscr{U}_1 = \mathscr{W}(\delta_1).$$

(12.12) COROLLARY. *Let $\mathscr{U}_i (i = 1, 2)$ be uniformities on X and Y respectively, $\mathscr{U}_2$ be totally bounded and $\delta_i = \delta(\mathscr{U}_i)$. Then $f: X \to Y$ is a proximity mapping if and only if it is uniformly continuous.*

(12.13) LEMMA. *If $\mathscr{U} \in \Pi(\delta)$ and $\mathscr{V}$ is a uniformity such that $\mathscr{W}(\delta) \subset \mathscr{V} \subset \mathscr{U}$, then $\mathscr{V} \in \Pi(\delta)$.*

Proof. This result follows immediately from (10.7), as

$$\delta < \delta(\mathscr{V}) < \delta.$$

(12.14) THEOREM. *If $\mathscr{U} \in \Pi(\delta)$, then $\mathscr{W} = \mathscr{W}(\delta)$ is the largest totally bounded uniformity contained in $\mathscr{U}$.*

Proof. If $\mathscr{U}^*$ is any totally bounded uniformity contained in $\mathscr{U}$, then $\mathscr{U}^* \vee \mathscr{W}$ is totally bounded and $\mathscr{W} \subset \mathscr{U}^* \vee \mathscr{W} \subset \mathscr{U}$. By (12.13), $\mathscr{U}^* \vee \mathscr{W} \in \Pi(\delta)$ and by (12.5), $\mathscr{U}^* \vee \mathscr{W} = \mathscr{W}$; that is, $\mathscr{U}^* \subset \mathscr{W}$.

(12.15) REMARKS. The above discussion establishes a one-to-one correspondence between all proximity structures and all totally bounded uniform structures. Since a completion of a totally bounded uniform space is a compact completely regular space, this then provides an alternate approach to the construction of the Smirnov compactification of a proximity space.

The following example shows that, in general, $\Pi(\delta)$ does not possess a largest member.

(12.16) EXAMPLE. Let X be a set admitting two countable partitions $\{A_i : i \in N\}$ and $\{B_j : j \in N\}$ such that $A_i \cap B_j \neq \varnothing$ for

$i, j \in N$. Let $\mathcal{U}_1$ and $\mathcal{U}_2$ be the uniform structures generated by the single entourages $U_1 = \bigcup_{i \in N} (A_i \times A_i)$ and $U_2 = \bigcup_{j \in N} (B_j \times B_j)$ respectively. As usual, let $\mathcal{W}_1$ and $\mathcal{W}_2$ denote the respective totally bounded uniform structures in the proximity classes of $\mathcal{U}_1$ and $\mathcal{U}_2$.

If $\delta_1 = \delta(\mathcal{U}_1 \bigvee \mathcal{U}_2)$ and $\delta_2 = \delta(\mathcal{W}_1 \bigvee \mathcal{W}_2)$, then we shall show that $\delta_2 \lneq \delta_1$. Clearly $\delta_2 < \delta_1$. Now $\mathcal{U}_1 \bigvee \mathcal{U}_2$ is defined by the single entourage

$$U_1 \cap U_2 = \bigcup_{i, j \in N} [(A_i \cap B_j) \times (A_i \cap B_j)],$$

and $\mathcal{W}_1 \bigvee \mathcal{W}_2$ has a base of entourages of the form

$$W = \bigcup_{\substack{m = 1, \dots, M_1 \\ n = 1, \dots, M_2}} [(E_m \cap F_n) \times (E_m \cap F_n)],$$

where $\{E_m\}$ and $\{F_n\}$ are finite partitions of X such that each E_m is a union of sets A_i, and each F_n is a union of sets B_j. Define

$$G = \bigcup_{k \in N} \left[\left(\bigcup_{l=1}^{k} A_l \right) \cap B_k \right].$$

Then $G \delta_1 (X - G)$. However, $G \delta_2 (X - G)$ as the following argument shows. Let $W \in \mathcal{W}_1 \bigvee \mathcal{W}_2$. Among the sets E_m, there exists at least one E_{m_0} which contains more than one A_i. Suppose

$$A_{i_1} \cup A_{i_2} \subset E_{m_0}, \quad \text{where} \quad i_1 < i_2.$$

Among the sets F_n, there exists a unique F_{n_0} containing B_{i_1}. Let $x \in A_{i_1} \cap B_{i_1}$ and $y \in A_{i_2} \cap B_{i_1}$. Then $x \in G$, $y \in (X - G)$ and

$$(x, y) \in [(E_{m_0} \cap F_{n_0}) \times (E_{m_0} \cap F_{n_0})] \subset W.$$

Since W is arbitrary, this shows that $G \delta_2 (X - G)$. Thus $\delta_2 \lneq \delta_1$.

We next show that $\Pi(\delta_2)$ does not have a largest member. Suppose on the contrary that $\mathcal{W}'$ is the largest member of $\Pi(\delta_2)$. Consider the uniform structures

$$\mathcal{V}_1 = \mathcal{U}_1 \bigvee \mathcal{W}_2 = \mathcal{U}_1 \bigvee (\mathcal{W}_1 \bigvee \mathcal{W}_2)$$

and

$$\mathcal{V}_2 = \mathcal{U}_2 \bigvee \mathcal{W}_1 = \mathcal{U}_2 \bigvee (\mathcal{W}_1 \bigvee \mathcal{W}_2).$$

Clearly $\delta(\mathcal{V}_1) < \delta_2$ and $\delta(\mathcal{V}_2) < \delta_2$ (as can be seen from the fact that $\delta(\mathcal{U}_1) = \delta(\mathcal{W}_1) < \delta(\mathcal{W}_1 \bigvee \mathcal{W}_2) = \delta_2$, and by (12.8),

$$\mathcal{V}_1, \mathcal{V}_2 \in \Pi(\delta_2).$$

Hence $\mathscr{V}_1 \subset \mathscr{W}'$ and $\mathscr{V}_2 \subset \mathscr{W}'$. Therefore $\mathscr{V}_1 \vee \mathscr{V}_2 \subset \mathscr{W}'$, i.e. $\mathscr{V}_1 \vee \mathscr{V}_2 \in \Pi(\delta_2)$. On the other hand, $\mathscr{V}_1 \vee \mathscr{V}_2 = \mathscr{U}_1 \vee \mathscr{U}_2 \in \Pi(\delta_1)$, a contradiction.

If δ is a pseudo-metric proximity, then $\Pi(\delta)$ does in fact have a largest member. In order to prove this, we first need a lemma:

(12.17) LEMMA. *Let U and W be symmetric subsets of $X \times X$ such that $\overset{4}{W} \subset U$. Then for every sequence $((x_n, y_n))$ from $X \times X$ such that $(x_n, y_n) \notin U$ for each $n \in N$, there exists a subsequence $((x_{n_k}, y_{n_k}))$ such that $(x_{n_k}, y_{n_l}) \notin W$ for every $k, l \in N$.*

Proof. For each $n \in N$, let $B_n = \{m : (x_n, y_m) \in W\}$ and

$$C_n = \{m : (x_m, y_n) \in W\}.$$

If $p, q \in B_n$, then $(y_p, y_q) \in \overset{2}{W}$. Hence $(x_q, y_p) \notin \overset{2}{W}$, since otherwise $(x_q, y_q) \in \overset{4}{W} \subset U$, contrary to hypothesis. Thus if B_n is infinite for any n, then $((x_m, y_m))_{m \in B_n}$ is the required subsequence. The case when C_n is infinite is similarly disposed of. The only remaining case is that for which both B_n and C_n are finite for each $n \in N$. Let $\phi(n)$ denote the first natural number greater than both n and all the elements of B_m and C_m $(m = 1, \ldots, n)$. Define $n_1 = 1$ and $n_{k+1} = \phi(n_k)$. Then $((x_{n_k}, y_{n_k}))$ is a subsequence with the desired property.

(12.18) THEOREM. *If a proximity δ is induced by a pseudo-metric d on X, then the corresponding pseudo-metric uniformity $\mathscr{U}$ is the largest member of $\Pi(\delta)$.*

Proof. If

$$U_n = \left\{ (x, y) : d(x, y) < \frac{1}{n} \right\},$$

then $\{U_n : n \in N\}$ is a base for $\mathscr{U}$. If the theorem is not true, there exists a $\mathscr{V} \in \Pi(\delta)$ such that $\mathscr{V} \not\subset \mathscr{U}$; that is, there exists a $V \in \mathscr{V}$ such that $\mathscr{U}_n \not\subset V$ for each $n \in N$. Thus for each $n \in N$, there exists a pair $(x_n, y_n) \in U_n - V$. Let W be a symmetric member of $\mathscr{V}$ such that $\overset{4}{W} \subset V$. Then by the previous lemma, there is a subsequence $((x_{n_k}, y_{n_k}))$ such that $(x_{n_k}, y_{n_l}) \notin W$ for every

$k, l \in N$. Set $A = \cup \{x_{n_k}\}$ and $B = \cup \{y_{n_k}\}$. Then $A \, \delta \, B$, since given any $n \in N$ there is a $k > n$ such that $x_{n_k} \in A$, $y_{n_k} \in B$ and

$$(x_{n_k}, y_{n_k}) \in U_{n_k} \subset U_n.$$

But $(A \times B) \cap W = \varnothing$ and therefore $A \, \hat{\delta} \, B$, a contradiction.

(12.19) COROLLARY. $\Pi(\delta)$ *contains at most one pseudo-metric uniformity. If it does contain one, then it is the largest element of the proximity class.*

(12.20) COROLLARY. *Let $\mathscr{U}_1$ be a pseudo-metric uniformity on X and let $\delta_1 = \delta(\mathscr{U}_1)$. Let δ_2 be a proximity on Y induced by a uniformity $\mathscr{U}_2$. Then $f \colon X \to Y$ is a proximity mapping if and only if it is uniformly continuous.*

13. Generalized uniform structures

In the previous section it was shown that if δ is a pseudo-metric proximity, then the corresponding pseudo-metric uniformity is the largest member of $\Pi(\delta)$. This naturally suggests the problem of characterizing those proximities δ which enjoy the property: $\Pi(\delta)$ has a largest member. In the present section this problem is studied using a generalized notion of uniform structure, which differs from the usual uniform structure in that a weaker intersection axiom is used (see 13.1 (*) below). *Total* uniform structures, whose role in the theory is dual to that of totally bounded structures, are also considered. Finally, *correct* structures are introduced, with the help of which a characterization of those filters from $X \times X$ which induce a separated proximity is obtained.

(13.1) DEFINITION. *An* Alfsen–Njåstad (*or* A–N) *uniform structure $\mathscr{U}$ on a set X has all the properties of a uniform structure* (10.1) *with the possible exception of* 10.1 (ii), *which is replaced by*:

(*)   *If $\{A_i \colon 1 \leqslant i \leqslant n\}$ is a family of subsets of X and $U_i \in \mathscr{U}$ for $1 \leqslant i \leqslant n$, then there is a $U \in \mathscr{U}$ such that*

$$U[A_i] \subset U_i[A_i], \quad 1 \leqslant i \leqslant n.$$

Members of $\mathscr{U}$ are called *entourages*, and the pair $(X, \mathscr{U})$ is called an A–N *uniform space*. $\mathscr{U}$ is said to be *separated* if it also satisfies 10.1 (vi).

Although it is evident that every uniform structure is an A–N uniform structure, it will be shown later that there exist A–N uniform structures which are not uniform structures. Many of the results of the previous sections concerning uniformities hold for A–N uniformities, and are summarized in the following theorem:

(13.2) THEOREM. *Every A–N uniform structure* $\mathcal{U}$ *induces a proximity structure* $\delta = \delta(\mathcal{U})$, *and hence a topology*

$$\tau = \tau(\delta) = \tau(\mathcal{U})$$

also, in the same way as a uniform structure. $\tau(\mathcal{U})$ *is completely regular and, in fact, Tychonoff if* $\mathcal{U}$ *is separated. If* $\Lambda(\delta)$ *denotes the proximity class of A–N uniform structures, then* $\Lambda(\delta)$ *contains a unique totally bounded member* $\mathcal{W}$, *and* $\mathcal{W}$ *is the smallest member of the class. Every totally bounded A–N uniform structure is a uniform structure.*

(13.3) DEFINITION. *A subset* V *of* $X \times X$ *is called* entourage-like *with respect to a proximity* δ *on* X *iff there exists a sequence* $(V_n)_{n \in N}$ *of symmetric subsets of* $X \times X$ *such that*

$$V_{n+1} \subset V_n, \quad \text{where} \quad V_0 = V, \quad n = 0, 1, \dots,$$

and $\qquad A \ll V_n[A] \quad \text{for all} \quad A \subset X.$

Equivalently, we sometimes say that V is entourage-like with respect to $\Lambda(\delta)$ or with respect to $\mathcal{U} \in \Lambda(\delta)$.

Clearly, every entourage of an A–N uniform structure is entourage-like.

(13.4) DEFINITION. *An A–N uniform structure* $\mathcal{U}$ *is said to be* total *iff every entourage-like set with respect to* $\delta(\mathcal{U})$ *is an entourage.*

In contrast to the situation with uniform structures, we have the following result concerning A–N uniform structures:

(13.5) THEOREM. *Every proximity equivalence class* $\Lambda(\delta)$ *contains a largest member* $\mathcal{A}$, *which consists of all entourage-like sets with respect to* δ.

Proof. Let $\mathcal{W}$ be the smallest member of $\Lambda(\delta)$ and let $\mathcal{A}$ be the class of all entourage-like sets relative to δ. Clearly $\mathcal{W} \subset \mathcal{A}$. To prove that $\mathcal{A}$ is an A–N uniform structure it suffices to verify 13.1(*), since the other axioms follow easily. Let

$\{A_i : 1 \leqslant i \leqslant n\}$ be a family of subsets of X and let $U_i \in \mathscr{A}$ for $1 \leqslant i \leqslant n$. Now by (13.3), $A_i \ll U_i[A_i]$ for $1 \leqslant i \leqslant n$. Since $\mathscr{W} \in \Lambda(\delta)$, there exist $W_i \in \mathscr{W}$ such that $W_i[A_i] \subset U_i[A_i]$ for $1 \leqslant i \leqslant n$. Since $\mathscr{W}$ is a uniform structure, $W = \bigcap_{i=1}^{n} W_i \in \mathscr{W} \subset \mathscr{A}$, and $W[A_i] \subset U_i[A_i]$ for $1 \leqslant i \leqslant n$.

Clearly $\delta < \delta(\mathscr{A})$ and, by (13.3), $\delta(\mathscr{A}) < \delta$; i.e. $\mathscr{A} \in \Lambda(\delta)$. Since every entourage of $\mathscr{U} \in \Lambda(\delta)$ is entourage-like relative to δ, $\mathscr{A}$ is the largest member of $\Lambda(\delta)$.

(13.6) COROLLARY. *In the above theorem, $\mathscr{A}$ is the only total structure of $\Lambda(\delta)$.*

(13.7) THEOREM. *The largest A–N uniform structure $\mathscr{A} \in \Lambda(\delta)$ is equal to each of the following*:

(i) *the union $\mathscr{U}_1$ of all members of $\Lambda(\delta)$.*
(ii) *the union $\mathscr{U}_2$ of all members of the classes $\Lambda(\delta')$ for all $\delta' < \delta$.*
(iii) *the union $\mathscr{U}_3$ of all members of $\Pi(\delta)$.*
(iv) *the union $\mathscr{U}_4$ of all members of the classes $\Pi(\delta')$ for all $\delta' < \delta$.*
(v) *the union $\mathscr{U}_5$ of all pseudo-metric members of the classes $\Pi(\delta')$, for all $\delta' < \delta$.*

Proof. Since $\mathscr{A} \in \Lambda(\delta)$, $\mathscr{A} \subset \mathscr{U}_1$. We also have $\mathscr{U}_2 \subset \mathscr{A}$ since every entourage of a member of $\Lambda(\delta')$, with $\delta' < \delta$, is entourage-like relative to δ. Thus $\mathscr{A} \subset \mathscr{U}_1 \subset \mathscr{U}_2 \subset \mathscr{A}$, from which (i) and (ii) follow.

Statements (iv) and (v) are implied by the relations

$$\mathscr{A} \subset \mathscr{U}_5 \subset \mathscr{U}_4 \subset \mathscr{A}.$$

The non-trivial part $\mathscr{A} \subset \mathscr{U}_5$ is due to the fact that, if V is an entourage-like set with a defining sequence (V_n) as in (13.3), then $\{V_n\}$ is a base for a pseudo-metric uniformity which contains V. (Recall that a uniformity with a countable base is pseudo-metrizable.)

Finally, to prove (iii), it suffices to show that $\mathscr{A} \subset \mathscr{U}_3$ since the reverse inclusion is immediate. We know from (iv) that if

$V \in \mathscr{A}$, then there exists a $\mathscr{V} \in \Pi(\delta')$ with $\delta' < \delta$ and such that $V \in \mathscr{V}$. Clearly

$$V \in \mathscr{V} \vee \mathscr{W}(\delta), \quad \text{and by (12.8),} \quad \mathscr{V} \vee \mathscr{W}(\delta) \in \Pi(\delta).$$

(13.8) COROLLARY. *If $\Pi(\delta)$ has a largest member, then such a member is precisely $\mathscr{A}$.*

The next theorem, which follows from the above discussion, characterizes those $\Pi(\delta)$'s which possess a largest member.

(13.9) THEOREM. *$\Pi(\delta)$ has a largest member if and only if the entourage-like sets with respect to δ form a filter.*

(13.10) REMARKS. (i) Example (12.16) illustrates the existence of a total A–N uniform structure which is not a uniform structure.

(ii) Theorem (12.18) illustrates that every pseudo-metric uniformity is total.

(13.11) DEFINITION. *A contiguity $\mathscr{U}$ on X is a family of symmetric and reflexive relations such that if $U \in \mathscr{U}$ and $U \subset V$ where $\Delta \subset V = V^{-1}$, then $V \in \mathscr{U}$.*

Given a contiguity $\mathscr{U}$ on X, one can define a binary relation $\delta = \delta(\mathscr{U})$ on the power set of X as follows:

(13.12) $A \delta B$ iff there is a $U \in \mathscr{U}$ such that $U[A] \cap B = \varnothing$.

A natural question arises as to what conditions are necessary and sufficient to ensure that $\delta = \delta(\mathscr{U})$ is a separated proximity on X.

(13.13) DEFINITION. *A contiguity $\mathscr{U}$ on X is called correct iff $\delta = \delta(\mathscr{U})$ is a separated proximity on X.*

We have seen that separated uniform and A–N uniform structures are correct. The following theorem gives a characterization of correct spaces:

(13.14) THEOREM. *A contiguity $\mathscr{U}$ on X is correct if and only if each of the following conditions is satisfied:*
   (i) *If $x, y \in X$, then $x \neq y$ iff there is a $U \in \mathscr{U}$ such that*

$$(x, y) \notin U.$$

(ii) *For each $A \subset X$ and $U, V \in \mathcal{U}$, there exists a $W \in \mathcal{U}$ such that*
$$W[A] \subset U[A] \cap V[A].$$

(iii) *For $A \subset X$ and $U \in \mathcal{U}$, there exist V, $W \in \mathcal{U}$ such that* $W[V[A]] \subset U[A]$.

Proof. Let us first suppose $\mathcal{U}$ to be correct and let $\delta = \delta(\mathcal{U})$. We need only prove (ii) and (iii). If (ii) is not satisfied, there exists an $A \subset X$ and U, $V \in \mathcal{U}$ such that for each $W \in \mathcal{U}$, we can find an $x = x(W) \in W[A] - (U[A] \cap V[A])$. Consider

$$B = \{x = x(W) : W \in \mathcal{U}\}.$$

Then $B = B_1 \cup B_2$ where $U[A] \cap B_1 = \varnothing$ and $V[A] \cap B_2 = \varnothing$. As this implies $A \not{\delta} B_1$ and $A \not{\delta} B_2$, we have $A \not{\delta} B$, contradicting the fact that $W[A] \cap B \neq \varnothing$ for each $W \in \mathcal{U}$.

To prove (iii), let $A \subset X$ and $U \in \mathcal{U}$. Since

$$U[A] \cap (X - U[A]) = \varnothing, \quad A \not{\delta} (X - U[A]).$$

By (3.5) (which is equivalent to the Strong Axiom), there exist disjoint δ-neighbourhoods E and F of A and $(X - U[A])$ respectively. Then $A \not{\delta} (X - E)$ and $(X - U[A]) \not{\delta} (X - F)$, implying the existence of V, $W \in \mathcal{U}$ such that $V[A] \cap (X - E) = \varnothing$ and

$$W[X - U[A]] \cap (X - F) = \varnothing.$$

Hence $V[A] \subset E$ and $W[X - U[A]] \subset F$. But $E \cap F = \varnothing$, so that

$$V[A] \cap W[X - U[A]] = \varnothing.$$

Therefore $W[V[A]] \cap (X - U[A]) = \varnothing$, since $W = W^{-1}$, which proves (iii).

To prove the converse, it is sufficient to verify the Strong Axiom. If $A \not{\delta} B$, then there is a $U \in \mathcal{U}$ such that $U[A] \cap B = \varnothing$. By 13.14(iii), there exist V, $W \in \mathcal{U}$ such that $W[V[A]] \subset U[A]$. We first verify that $V[A] \cap W[B] = \varnothing$. Suppose instead that there exists an $x \in V[A] \cap W[B]$. Then $W[x] \subset W[V[A]] \subset U[A]$ and there exists a $y \in W[x] \cap B \subset U[A]$. But this contradicts the fact that $U[A] \cap B = \varnothing$. Clearly $V[A]$ and $W[B]$ are disjoint δ-neighbourhoods of A and B respectively.

An example is now given of a correct space which is not a uniform space:

(13.15) EXAMPLE. Let an increasing sequence (a_n) of natural numbers be called *rapidly increasing* iff $a_n/n \to \infty$. Let X be the set of all natural numbers and $\mathcal{U}'$ be the collection of all sequences which either have a finite range or are rapidly increasing. For each $U' \in \mathcal{U}'$ define

$$U = \{(x,y) \in X \times X : \text{either } x \text{ and } y \text{ are both in or} \\ \text{both not in the range of } U'\}.$$

Then $\mathcal{U} = \{U : U' \in \mathcal{U}'\}$ is a contiguity on X.

For $M \subset X$ and $U \in \mathcal{U}$, let us first determine $U[M]$. Let N denote the range of U'. We are then faced with one of the three following situations:

(a) If $M \cap N = \varnothing$, then $U[M] = X - N$.

(b) If $M \subset N$, then $U[M] = N$.

(c) If $M \cap N \neq \varnothing$ and $M \not\subset N$, then $U[M] = X$.

The correctness of $\mathcal{U}$ will now be verified. If $x \neq y$, we take U corresponding to $N = \{x\}$. Then $y \notin U[x]$. Conversely, if $y \notin U[x]$, then $x \neq y$, verifying 13.14(i). For each $A \subset X$ and $U \in \mathcal{U}$, $U[U[A]] = U[A]$ and 13.14(iii) is verified. Finally, to show 13.14(ii), let $A \subset X$ and $U, V \in \mathcal{U}$. Let N_1 and N_2 denote the respective ranges of the sequences U' and V'.

We now consider the three possible situations separately:

(I) If $A \cap N_1 = \varnothing = A \cap N_2$, then $U[A] = X - N_1$ and $V[A] = X - N_2$. Let $N_0 = N_1 \cup N_2$. Then $N_0 \in U'$ and, if W is the corresponding member of $\mathcal{U}$, then

$$W[A] \subset X - N_0 \subset U[A] \cap V[A].$$

(II) If $A \subset N_1$, then A is the range of some W'. In this case, $W[A] = A \subset U[A] \cap V[A]$.

(III) If $A \cap N_1 \neq \varnothing$ and $A \not\subset N_1$, then $U[A] = X$ and we may choose $W = V$.

Finally, $\mathcal{U}$ is shown not to be a uniformity by observing that 10.1(ii) is not satisfied. Suppose $x \neq y$ and let U, V be members of $\mathcal{U}$ corresponding to $U' = (x)$ and $V' = (y)$. We then clearly have $U \cap V \notin \mathcal{U}$.

14. Proximity and height

In this section, a new partial order for uniform structures is introduced through the concept of *height*. In some respects, this concept is dual to that of proximity and will clarify the order structure of $\Pi(\delta)$.

(14.1) DEFINITION. *A subset U of $X \times X$ is* totally bounded *iff there exists a family of sets $\{A_i : 1 \leqslant i \leqslant n\}$ such that $X = \bigcup_{i=1}^{n} A_i$ and $A_i \times A_i \subset U$ for $1 \leqslant i \leqslant n$. A uniform structure is* totally bounded *iff every entourage is totally bounded. This is equivalent to the condition: for each $U \in \mathscr{U}$ there exists a finite set*

$$\{x_1, \ldots, x_n\} \subset X$$

such that $X = \bigcup_{i=1}^{n} U[x_i]$ (sometimes called precompactness).

(14.2) DEFINITION. *Let $\mathscr{U}$ and $\mathscr{V}$ be two uniformities on X. $\mathscr{U}$ is said to be* less than or equal in height *($\leqslant$) to $\mathscr{V}$ iff for each $U \in \mathscr{U}$, there exists a $V \in \mathscr{V}$ such that $U \cup V^c$ is totally bounded (where $V^c = X \times X - V$).*

Equivalently, we may define $\mathscr{U} \leqslant \mathscr{V}$ iff for each $U \in \mathscr{U}$, there exists a $V \in \mathscr{V}$ and a finite family $\{A_i : 1 \leqslant i \leqslant n\}$ with

$$\bigcup_{i=1}^{n} A_i = X \quad \text{such that} \quad V \cap (A_i \times A_i) \subset U \quad \text{for each } i.$$

The following results are immediate consequences of this second form of the definition:

(14.3) LEMMA. *The relation $\leqslant$ as defined above is a preorder (i.e. reflexive and transitive) on the set of all uniformities on X.*

(14.4) LEMMA. *If $\mathscr{U} \subset \mathscr{V}$, then $\mathscr{U} \leqslant \mathscr{V}$.*

Clearly $\leqslant$ induces an equivalence relation on the set of all uniformities on X, and so gives rise to equivalence classes. We shall denote by $H(\mathscr{U})$, called the *height class of $\mathscr{U}$*, the collection of all uniformities on X which are equal in height to $\mathscr{U}$. The following result is obvious.

(14.5) LEMMA. *The smallest height class is the class of all totally bounded uniformities on X.*

The next theorem provides the first step towards showing that the family of all height classes of uniformities on a set X is a complete lattice. The corresponding result for the family of all uniformities, assigned the partial order induced by set inclusion, is well known.

(14.6). THEOREM. *If $\mathscr{U} \leqslant \mathscr{V}$ and $\mathscr{W} \leqslant \mathscr{V}$, then $(\mathscr{U} \vee \mathscr{W}) \leqslant \mathscr{V}$.*

Proof. A typical member of $\mathscr{U} \vee \mathscr{W}$ is $U \cap W$, where U and W are members of $\mathscr{U}$ and $\mathscr{W}$ respectively. By hypothesis, there exist entourages $V_1, V_2 \in \mathscr{V}$ and finite covers $\{A_i : 1 \leqslant i \leqslant n\}$ and $\{B_j : 1 \leqslant j \leqslant m\}$ of X such that $A_i \times A_i \subset U \cup V_1^c$ and

$$B_j \times B_j \subset W \cup V_2^c.$$

Clearly $\{A_i \cap B_j : 1 \leqslant i \leqslant n, 1 \leqslant j \leqslant m\}$ is a finite cover of X and $(A_i \cap B_j) \times (A_i \cap B_j) \subset (U \cap W) \cup (V_1 \cap V_2)^c$, proving the theorem.

The following corollary is an immediate consequence of the above theorem together with (14.4).

(14.7) COROLLARY. *If $\mathscr{U} \leqslant \mathscr{V}$, then $(\mathscr{U} \vee \mathscr{V}) \in H(\mathscr{V})$.*

(14.8) COROLLARY. *If $\mathscr{U}$ is totally bounded, then*

$$(\mathscr{U} \vee \mathscr{V}) \in H(\mathscr{V})$$

for every $\mathscr{V}$.

NOTATION. If $\mathscr{U}$ is a uniformity on X and $\delta = \delta(\mathscr{U})$, we shall denote by $\mathscr{U}_w$ the unique totally bounded member of $\Pi(\delta)$. Recall that such a member exists by Corollaries (12.4) and (12.5).

(14.9) THEOREM. *If $\mathscr{U} \leqslant \mathscr{V}$, then $\mathscr{U} \subset [\mathscr{V} \vee (\mathscr{U} \vee \mathscr{V})_w]$.*

Proof. To simplify the proof, we suppose that all entourages throughout the discussion are symmetric.

Given $U \in \mathscr{U}$, let $U_1, U_2 \in \mathscr{U}$ be such that $\overset{3}{U_1} \subset U$ and $\overset{3}{U_2} \subset U_1$. Since $\mathscr{U} \leqslant \mathscr{V}$, there exists a $V \in \mathscr{V}$ and a finite cover

$$\{A_i : 1 \leqslant i \leqslant n\}$$

of X such that $\displaystyle\bigcup_{i=1}^{n} A_i \times A_i \subset U_1 \cup V^c.$

Let $V_1 \in \mathscr{V}$ be such that $\overset{3}{V_1} \subset V$, and define $B_i = (U_2 \cap V_1)[A_i]$. Then $\{B_i : 1 \leqslant i \leqslant n\}$ is a $\delta(\mathscr{U} \vee \mathscr{V})$-cover of X, so that

$$U_w = \bigcup_{i=1}^{n} (B_i \times B_i) \in (\mathscr{U} \vee \mathscr{V})_w.$$

The theorem is now proved, since $U_w \cap V_1 \subset U$ as verified below. If $(t, u) \in U_w \cap V_1$, then both t and u belong to B_i for some i; that is, there exist $x, y \in A_i$ such that $(x, t), (y, u) \in U_2 \cap V_1$. Hence

$$(x, y) \in V \quad \text{and, since} \quad A_i \times A_i \subset U_1 \cup V^c,$$

$(x, y) \in U_1$. Since $(x, t), (y, u) \in U_2 \subset U_1$ and all entourages are symmetric, it follows that $(t, u) \in U$, as required.

(14.10) COROLLARY. *If $\Pi(\delta)$ has a least upper bound, then $\mathscr{U} \leqslant \mathscr{V}$ for $\mathscr{U}, \mathscr{V} \in \Pi(\delta)$ implies that $\mathscr{U} \subset \mathscr{V}$.*

Proof. If $\mathscr{W}$ is the least upper bound of $\Pi(\delta)$, then

$$\mathscr{W}_w = (\mathscr{U} \vee \mathscr{V})_w \subset \mathscr{V}.$$

Hence $\mathscr{V} = [\mathscr{V} \vee (\mathscr{U} \vee \mathscr{V})_w]$ and, by (14.9),

$$\mathscr{U} \subset [\mathscr{V} \vee (\mathscr{U} \vee \mathscr{V})_w]_i = \mathscr{V}.$$

No confusion should arise from the fact that the same notation $\leqslant$ is used for the preorder on the set of all uniformities on X, as for the preorder which it induces on the height classes.

(14.11) COROLLARY. *If H_1 and H_2 are two height classes, then $H_1 \leqslant H_2$ iff there exist a $\mathscr{U}_1 \in H_1$ and a $\mathscr{U}_2 \in H_2$ such that $\mathscr{U}_1 \subset \mathscr{U}_2$.*

Proof. In view of Lemma (14.4), it is sufficient to prove necessity. Let $H_1 \leqslant H_2$, $\mathscr{U}_1 \in H_1$ and $\mathscr{V} \in H_2$. Then $\mathscr{U}_1 \leqslant \mathscr{V}$ and, by (14.5) and (14.7),

$$\mathscr{U}_2 = [\mathscr{V} \vee (\mathscr{U}_1 \vee \mathscr{V})_w] \in H_2.$$

According to Theorem (14.9), $\mathscr{U}_1 \subset \mathscr{U}_2$.

In Corollaries (12.4) and (12.5), we have seen that every $\Pi(\delta)$ has a smallest member. The following result concerning the height classes is, in a sense, dual to this.

(14.12) THEOREM. *Every height class H has a largest member $\mathscr{U}_h$. If $\mathscr{U} \leqslant \mathscr{U}_h$, then $\mathscr{U} \subset \mathscr{U}_h$.*

Proof. Let $\mathscr{U}_\Delta$ be the largest uniformity on X: namely, that generated by the diagonal Δ. Then every finite cover of X is a $\delta(\mathscr{U}_\Delta)$-cover of X. For each $\mathscr{U} \in H$, define

$$\mathscr{U}_h = [\mathscr{U} \vee (\mathscr{U}_\Delta)_w].$$

Then $\mathscr{U}_h \in H$ and, if $\mathscr{V} \leqslant \mathscr{U}_h$, Theorem (14.9) implies that

$$\mathscr{V} \subset [\mathscr{U}_h \vee (\mathscr{V} \vee \mathscr{U}_h)_w];$$

that is, $\mathscr{V} \subset \mathscr{U}_h$.

(14.13) COROLLARY. *The family of all height classes of uniformities on a set X forms a complete lattice.*

Proof. In view of (14.5), it suffices to show that every subfamily $\{H_\alpha : \alpha \in A\}$ of height classes has a least upper bound. Let $\mathscr{U}_\alpha$ be the largest member of H_α for each $\alpha \in A$, and let $\mathscr{U} = \bigvee_{\alpha \in A} \mathscr{U}_\alpha$. Then $H(\mathscr{U})$ is an upper bound for $\{H_\alpha : \alpha \in A\}$. If H' is any upper bound for $\{H_\alpha : \alpha \in A\}$ and $\mathscr{V}_h$ is the largest member of H', then (14.12) implies that $\mathscr{U}_\alpha \subset \mathscr{V}_h$ for each $\alpha \in A$. Hence $\mathscr{U} \subset \mathscr{V}_h$ and, by (14.11), $H(\mathscr{U}) \leqslant H'$.

15. Hyperspace uniformities

If (X, τ) is a uniformizable space, then any compatible uniformity $\mathscr{U}$ on X induces a uniformity $\mathbf{\mathscr{U}}$ on the hyperspace $H(X)$ of all non-empty closed subsets of X as follows: for each $U \in \mathscr{U}$, an element $\mathbf{U} \in \mathbf{\mathscr{U}}$ is defined by

$$\mathbf{U} = \{(A, B) \in H(X) \times H(X) : A \subset U[B] \quad \text{and} \quad B \subset U[A]\}.$$

Two compatible uniformities on X which induce compatible uniformities on $H(X)$ will be referred to as *H-equivalent*, examples of which can be found in the literature. The following questions naturally arise:

(a) Under what conditions are two uniformities H-equivalent?

(b) Under what conditions does H-equivalence of uniformities imply identity?

There is an ever-growing literature centred on these problems (see Notes for references); in this section, we merely study the role played by proximity in such problems.

(15.1) THEOREM. *If two uniformities on X are H-equivalent, then they are in the same proximity class.*

Proof. Suppose $\mathscr{U}_1$ and $\mathscr{U}_2$ are two uniformities on X which are not in the same proximity class; that is, if $\delta_i = \delta(\mathscr{U}_i)$ for $i = 1, 2$, then $\delta_1 \neq \delta_2$. Then by (2.8), there are closed subsets A and B of X such that $A\,\delta_1\,B$ but $A\,\delta_2\,B$. In other words, $U_1[B] \cap A \neq \varnothing$ for each $U_1 \in \mathscr{U}_1$, while there exists a $U_2 \in \mathscr{U}_2$ such that $U_2[B] \cap A = \varnothing$. Let $\mathbf{A} = \{F \in H(X) : F \cap A \neq \varnothing\}$. Given $U_1 \in \mathscr{U}_1$, there exists a $V_1 \in \mathscr{U}_1$ such that $V_1 \circ V_1 \subset U_1$. Choose $a \in V_1[B] \cap A$. Then

$$\overline{\{a\}} \subset V_1[a] \subset V_1 \circ V_1[B] \subset U_1[B],$$

so that $B \cup \overline{\{a\}} \subset U_1[B]$. Clearly $B \subset U_1[B \cup \overline{\{a\}}]$ and $B \cup \overline{\{a\}} \in \mathbf{A}$. Hence $\mathbf{U}_1[B] \cap \mathbf{A} \neq \varnothing$, and we conclude that $B \in \tau(\mathscr{U}_1)$-closure of $\mathbf{A}$. On the other hand, since $U_2[B]$ contains no set intersecting A, we know that $B \notin \tau(\mathscr{U}_2)$-closure of $\mathbf{A}$. Thus $\mathscr{U}_1$ and $\mathscr{U}_2$ are not H-equivalent, establishing the theorem.

(15.2) COROLLARY. *Two different uniformities on X, at least one of which is totally bounded, are not H-equivalent.*

Proof. This follows from the two facts:

(i) every proximity class contains exactly one totally bounded uniformity, by Corollary (12.5), and

(ii) the family of closures of finite subsets of X is dense in $(H(X)), \tau(\mathscr{U}))$ iff $\mathscr{U}$ is totally bounded.

To see the validity of statement (ii), suppose $U \in \mathscr{U}$ and let B be a finite subset of X such that $\bar{B} \in \mathbf{U}[X]$. Then clearly $X = U[B]$, so that $\mathscr{U}$ is totally bounded.

Conversely, if $\mathscr{U}$ is totally bounded and $U = U^{-1} \in \mathscr{U}$, then by (14.1) there exists a finite subset $\{x_1, \ldots, x_n\}$ of X such that $X = \bigcup_{i=1}^{n} U[x_i]$. Clearly, $\overline{\{x_i\}} \subset U[x_i]$ for $1 \leqslant i \leqslant n$. Given $A \in H(X)$, let $B = \{x_i : A \cap U[x_i] \neq \varnothing\}$. Then $A \subset U[\bar{B}]$ and $\bar{B} \subset U[A]$, so that $\bar{B} \in \mathbf{U}[A]$. Hence the closures of finite subsets of X are dense in $H(X)$.

(15.3) COROLLARY. *Two different uniformities on X, each having a countable base, are not H-equivalent.*

Proof. This follows from Theorem (12.18) and the fact that every uniform structure which has a countable base is pseudometrizable.

(15.4) DEFINITION. *Let $\mathscr{U}$ and $\mathscr{V}$ be two uniformities on X and let $A \subset X$. $\mathscr{U}$ is said to be* uniformly finer *than $\mathscr{V}$ on A over X iff given any $V \in \mathscr{V}$, there exists a $U \in \mathscr{U}$ such that $U \cap (A \times X) \subset V$.*

(15.5) DEFINITION. *$\mathscr{U}$ is* H-finer *than $\mathscr{V}$ iff $\tau(\mathscr{U}) \supset \tau(\mathscr{V})$ in $H(X)$.*

(15.6) DEFINITION. *A subset E of X is* U-discrete *for $U \in \mathscr{U}$ iff for each $x \in E$, $U[x] \cap E = \{x\}$. E is $\mathscr{U}$-discrete if it is U-discrete for some $U \in \mathscr{U}$.*

The next result answers question (*a*) of the introductory paragraph.

(15.7) THEOREM. *Let $\mathscr{U}$ and $\mathscr{V}$ be two uniformities on X. Then $\mathscr{U}$ is H-finer than $\mathscr{V}$ if and only if it is both* (i) *proximally finer, and* (ii) *uniformly finer over X on every $\mathscr{V}$-discrete set.*

Proof. Let $\mathscr{U}$ be *H*-finer than $\mathscr{V}$. Statement (i) is implicit in the proof of Theorem (15.1). In proving (ii), we again work solely with symmetric entourages. Suppose $E_0 \subset X$ is V_0-discrete for some $V_0 \in \mathscr{V}$. Let $V_1 \in \mathscr{V}$ and $V_2 = V_1 \cap V_0$. Since $\mathscr{U}$ is *H*-finer than $\mathscr{V}$, there is a $U \in \mathscr{U}$ such that $E \in \mathbf{U}[E_0]$ implies $E \in \mathbf{V}_2[E_0]$. In particular, let $E = \{y\} \cup (E_0 - \{x_0\})$ where $x_0 \in E_0$ and $(x_0, y) \in U$. Then $E_0 \subset V_2[E]$; that is, there exists a $y' \in E$ such that $(x_0, y') \in V_2 \subset V_0$. Since E_0 is V_0-discrete, we have $y' = y$ and hence

$$U \cap (E_0 \times X) \subset V_1.$$

To establish the converse, suppose that (i) and (ii) are satisfied, and let $E_0 \subset X$ and $V_0 \in \mathscr{V}$. Now E_0 and $X - V_0[E_0]$ are not near relative to $\delta(\mathscr{V})$, and hence not near relative to $\delta(\mathscr{U})$; that is, there is a $U_0 \in \mathscr{U}$ such that $U_0[E_0] \subset V_0[E_0]$. Let $V_1 \in \mathscr{V}$ be such that $\overset{2}{V_1} \subset V_0$ and let E_1 be a maximal V_1-discrete subset of E_0, so that $E_0 \subset V_1[E_1]$. By (ii), there exists a $U_1 \in \mathscr{U}$ such that

$$U_1 \cap (E_1 \times X) \subset V_1$$

and also, $U_1 \subset U_0$. If $E \subset X$ satisfies $E_0 \subset U_1[E]$, then for $x \in E_1$ there exists a $y \in E$ such that $(x, y) \in U_1$. Hence $(x, y) \in V_1$ and we have $E_1 \subset V_1[E]$. Therefore $E_0 \subset \overset{2}{V_1}[E] \subset V_0[E]$. Thus $E_0 \subset U_1[E]$ implies $E_0 \subset V_0[E]$ and, since $U_1 \subset U_0$,

$$E \subset U_1[E_0] \quad \text{implies} \quad E \subset V_0[E_0].$$

This proves that $\mathscr{U}$ is H-finer than $\mathscr{V}$.

(15.8) REMARK. The sufficiency of the above theorem is used in the next result in the following form: Let $\mathscr{U}$, $\mathscr{V}$ be two uniformities on X such that $\delta(\mathscr{V}) < \delta(\mathscr{U})$. If for every $V \in \mathscr{V}$ and every V-discrete set $A \subset X$ there exists a $U \in \mathscr{U}$ such that

$$U[a] \subset V[a] \quad \text{for all} \quad a \in A,$$

then $\mathscr{U}$ is H-finer than $\mathscr{V}$.

(15.9) THEOREM. *Let $\mathscr{U}$ and $\mathscr{V}$ be two uniformities on X which are equal in height and are in the same proximity class, i.e.*

$$\delta(\mathscr{U}) = \delta(\mathscr{V}).$$

Then $\mathscr{U}$ and $\mathscr{V}$ are H-equivalent.

Proof. Let $S \in \mathscr{V}$ and $D_1, \ldots, D_n$ be a finite cover of $A \subset X$. Since $\mathscr{U}$ and $\mathscr{V}$ are in the same proximity class, there exists a $U \in \mathscr{U}$ such that $U[D_i] \subset S[D_i]$ for all $i = 1, \ldots, n$. If $a \in A$ and $(a, x) \in U$, then

$$(a, x) \in \bigcup_{i=1}^{n} (D_i \times U[D_i]) \subset \bigcup_{i=1}^{n} (D_i \times S[D_i]).$$

Thus, in view of (15.8), it is sufficient to prove that given a $V \in \mathscr{V}$ and a V-discrete subset A of X, there exist $T \in \mathscr{U}$, $S \in \mathscr{V}$ and a finite cover $D_1, \ldots, D_n$ of A such that

$$(a, x) \in \bigcup_{i=1}^{n} (T \cap (D_i \times S[D_i]))$$

implies $(a, x) \in V$.

Let $W \in \mathscr{V}$ be such that $\overset{3}{W} \subset V$. Since $\mathscr{V} \leqslant \mathscr{U}$, there exists a finite cover $K_1, \ldots, K_r$ of X and a $T \in \mathscr{U}$ such that

$$\bigcup_{i=1}^{r} (\overset{3}{T} \cap (K_i \times K_i)) \subset W.$$

This in turn (since we also have $\mathscr{U} \leqslant \mathscr{V}$) yields a finite cover $E_1, \ldots, E_s$ of X and an $S \in \mathscr{V}$ such that $\overset{2}{S} \subset W$ and

$$\bigcup_{j=1}^{s} (\overset{2}{S} \cap (K_j \times K_j)) \subset T.$$

(i) Given an $x \in X$ there are at most r elements $a \in A$ such that $(a, x) \in T$. For if there were more than r, then at least two of them, say a_1 and a_2, would be members of the same K_i (for some i). Therefore $(a_1, a_2) \in (\overset{2}{T} \cap (K_i \times K_i)) \subset W \subset V$, a contradiction since A is V-discrete.

(ii) Given $a_0 \in A$, there exist at most r elements $a_k \neq a_0$ of A satisfying: for each k there exists an $x_k \in X$ such that

$$(a_0, x_k) \in T \quad \text{and} \quad (a_k, x_k) \in S.$$

If $x_k = x_j$ for $k \neq j$, then $(a_k, a_j) \in \overset{2}{S} \subset V$, a contradiction. Hence the x_k's are distinct. If there were more than r of them, at least two of them, say x_1 and x_2, would fall in the same K_i (for some i). In this case $(x_1, x_2) \in (\overset{2}{T} \cap (K_i \times K_i)) \subset W$, which in turn implies $(a_1, a_2) \in \overset{3}{W} \subset V$, a contradiction.

(iii) Given $a_0 \in A$, there exist at most sr^2 elements $a_k (\neq a_0)$ of A satisfying: for each k, there exists an $x_k \in X$ such that

$$(a_k, x_k) \in T \quad \text{and} \quad (a_0, x_k) \in S.$$

If this were not true, then by (i), more than sr of the x_k's would be distinct. Consequently, more than r of them, say $x_1, \ldots, x_{r+1}$ would fall in the same E_i (for some i) Hence

$$(x_k, x_j) \in (\overset{2}{S} \cap (E_i \times E_j)) \subset T$$

for all j, k between 1 and $r+1$. Thus for the same j's and k's,

$(a_k, a_j) \in \overset{3}{T}$. But at least two of these, say a_1 and a_2, are in the same K_i. Thus $(a_1, a_2) \in (\overset{3}{T} \cap (K_i \times K_i)) \subset W \subset V$, a contradiction.

Combining (ii) and (iii) we obtain:

(iv) Given $a_0 \in A$, there exist at most $n = r + sr^2$ members $a_k \neq a_0$ of A satisfying: for each k there exists an $x_k \in X$ such that either $(a_0, x_k) \in T$ and $(a_k, x_k) \in S$, or $(a_0, x_k) \in S$ and $(a_k, x_k) \in T$.

Now let $D_1, \ldots, D_{n+1}$ be maximal disjoint subsets of A such that for $a, a' \in D_i$ and any $x \in X$, $(a, x) \in T$ and $(a', x) \in S$ implies $a = a'$. By (iv), $A = \bigcup_{i=1}^{n+1} D_i$. Now $a \in A$, $x \in X$ and $(a, x) \in (T \cap (D_i \times S[D_i]))$ implies there exists an $a' \in D_i$ such that $(a', x) \in S$. But then $a = a'$ and $(a, x) \in S \subset V$, completing the proof.

Notes

10. The definition of a uniformity as given in this section is due to Weil [N]. That every such uniformity has a base consisting of open symmetric members is proved on p. 179 of Kelley [J]. Smirnov [98] was the first to investigate the relationships between proximity spaces and uniform spaces, although he worked with uniform structures defined by systems of uniform coverings. This equivalent method of defining a uniformity was earlier employed by Tukey [L]. Further work concerning the connection between uniformity and proximity was carried out by Gál [27] and by Alfsen and Fenstad [4] using Weil's uniform structures. Ramm gave the first example of two different uniform structures inducing the same proximity.

11. This section owes its existence to the work of the authors [121]. Definition (11.3) is equivalent to the usual one: namely, a uniform space is complete iff every Cauchy filter in the space converges to some point of the space.

12. A uniform structure $\mathcal{U}$ on X is *precompact* iff for every $U \in \mathcal{U}$, there are finitely many points $x_1, \ldots, x_n$ in X such that $X = \bigcup_{i=1}^{n} U[x_i]$. This concept is equivalent to that of *total boundedness* (see Gaal [C], p. 279) which is defined as follows: $(X, \mathcal{U})$ is totally bounded iff for every $U \in \mathcal{U}$, there is a finite cover $\{A_i : 1 \leq i \leq n\}$ of X such that $A_i \times A_i \subset U$ for each i. For more information regarding statements (i) and (ii) of the opening paragraph, the reader is referred to pp. 234–5 of Gillman and Jerison [E]. Statement (iii) was proved by Gál [D]. The equivalence of proximities

and totally bounded uniform structures has been shown by Smirnov [98], Alfsen and Fenstad [4] and Gál [27].

Our presentation of the theory in this section follows closely that of Thron [115] and of Alfsen and Njåstad [6]. Theorem (12.11) was independently proved by Alfsen–Njåstad [6] and Hursch (see Thron [115]). Example (12.16) was provided by Alfsen and Njåstad [6]. Another example illustrating the same point has been given by Leader [56]. Lemma (12.17) and Theorem (12.18) are due to Efremovič [19].

13. The first part of this section is based on the work of Alfsen and Njåstad [6]. Further interesting results concerning lattice operations, completions, etc., of generalized uniform structures are also given in this paper. In proving Theorem 13.7 (v), we make use of the fact that a uniform space with a countable base is pseudo-metrizable. The material on correct structures is due to Mordkovič [70] and Efremovič, Mordkovič and Sandberg [21]. The term 'contiguity' was defined by Lubkin in a somewhat different manner.

14. The concept of height was introduced by Hursch [37], who used the notation $\leqslant^h$ for the height relation. We have simply used the notation $\leqslant$, as there is no danger of confusion. This section is based on the papers of Hursch [37, 38]. Thron initially conjectured that there exists at most one uniformity in the intersection of a height class and a proximity class. This, however, was later disproved by Hursch. For details of this example along with other interesting results and examples, the reader is referred to the original papers.

15. It was first conjectured by Isbell [44] (p. 35, Ex. 17) that different uniformities on X are not H-equivalent; in fact, he suggested that they induce non-equivalent families of neighbourhoods of the element $X \in H(X)$. Smith [109] gave a counter-example to show the latter statement to be false, but nevertheless proved several results supporting Isbell's conjecture. Ward [M], however, disproved this conjecture. Theorem (15.1), along with Corollaries (15.2) and (15.3), is due to Smith [109]. Ward [118] proved Theorem (15.7), while Theorem (15.9) is due to Hursch [G]. The statement '$\mathscr{U}$ is proximally finer than $\mathscr{V}$' used in Theorem (15.7) means, of course, that $\delta(\mathscr{V}) < \delta(\mathscr{U})$. More material concerning this fascinating problem can be found in Isbell [I] and Ward [119].

CHAPTER 4

FURTHER DEVELOPMENTS

16. Proximal convergence

In this section we introduce the concept of *proximal convergence*, or *convergence in proximity*, which is analogous to uniform convergence. Proximal convergence preserves (proximal) continuity and enjoys interesting relationships with both uniform convergence and continuous convergence.

Recall that a *net* is a function on a directed set $(D, \geqslant)$. A subset E of D is a *cofinal* subset of $(D, \geqslant)$ iff for every $n \in D$, there exists an $m \in E$ such that $m \geqslant n$.

(16.1) DEFINITION. *Let (Y, δ) be a proximity space and let $(f_n : n \in D)$ be a net of functions mapping X to Y. Then (f_n) is said to* converge proximally *to $f \in Y^X$ iff $f(A)\,\delta\,B$ for $A \subset X$ and $B \subset Y$ implies $f_n(A)\,\delta\,B$ eventually.*

That uniform convergence is stronger than proximal convergence follows as a corollary to the next theorem.

(16.2) THEOREM. *Let $(Y, \mathscr{V})$ be an A–N uniform space and let $\delta = \delta(\mathscr{V})$. If a net (f_n) of mappings from X to Y converges uniformly to f, then the convergence is proximal.*

Proof. Let $A \subset X$, $B \subset Y$ and $f(A)\,\delta\,B$. Then there exists an entourage $V \in \mathscr{V}$ such that $\overset{2}{V}[f(A)] \cap B = \varnothing$. However, since (f_n) converges uniformly to $f, f_n(x) \in V[f(x)]$ eventually for each $x \in A$; that is, $f_n(A) \subset V[f(A)]$ eventually. We therefore eventually have $V[f_n(A)] \cap B = \varnothing$ and hence $f_n(A)\,\delta\,B$, showing that (f_n) converges to f proximally.

We now prove a near converse of the above result:

(16.3) THEOREM. *If a net $(f_n : n \in D)$ of mappings from X to a totally bounded uniform space $(Y, \mathscr{V})$ converges proximally to $f \in Y^X$, then the convergence is uniform.*

Proof. Suppose the convergence is not uniform with respect to $\mathscr{V}$. Then there exists a $V \in \mathscr{V}$, a cofinal subset D_o of D and, for each $n \in D_o$, an $x_n \in X$ such that $(f_n(x_n), f(x_n)) \notin V$. Choose $W \in \mathscr{V}$ such that $\overset{2}{W} \subset V$ and $W = \overset{m}{\underset{i=1}{\cup}} (A_i \times A_i)$, where $\{A_i : 1 \leqslant i \leqslant n\}$ is a δ-cover (see 12.6) of X. Now for some $i_0 \leqslant m$, there is a cofinal subset B of D_o such that $f_\beta(x_\beta) \in A_{i_0}$ for each $\beta \in B$. Set

$$G = \{x_\beta : \beta \in B\} \quad \text{and let} \quad x_p, x_q \in G.$$

If $(f_p(x_p), f(x_q)) \in W$, then $(f_q(x_q), f_p(x_p)) \in A_{i_0} \times A_{i_0} \subset W$ implies that $(f_q(x_q), f(x_q)) \in \overset{2}{W} \subset V$, a contradiction. Hence

$$(f_p(x_p), f(x_q)) \notin W, \quad \text{or} \quad f_p(x_p) \notin W[f(G)].$$

Therefore $f_\beta(G) \not\subset W[f(G)]$ for every $\beta \in B$, contradicting the fact that $(f_\beta : \beta \in B)$, and hence (f_n), converges proximally to f.

(16.4) COROLLARY. *If Y is a uniformizable space with a unique uniformity, then proximal and uniform convergence are equivalent.*

The following results show that proximal convergence preserves both continuity and proximal continuity.

(16.5) THEOREM. *Let (f_n) be a net of proximity mappings from (X, δ_1) to (Y, δ_2). If (f_n) converges proximally to f, then f is a proximity mapping.*

Proof. Suppose $A \delta_1 B$, but $f(A) \bar{\delta}_2 f(B)$. Then there is a subset E of Y such that $f(A) \bar{\delta}_2 E$ and $(Y - E) \bar{\delta}_2 f(B)$. We therefore eventually have $f_n(A) \bar{\delta}_2 E$ and $(Y - E) \bar{\delta}_2 f_n(B)$; that is, eventually $f_n(A) \bar{\delta}_2 f_n(B)$, contradicting the hypothesis that each f_n is proximally continuous.

The following result is proved similarly, using the characterization that a mapping f is continuous iff $f(\bar{A}) \subset \overline{f(A)}$ for each subset A of X.

(16.6) THEOREM. *If a net (f_n) of continuous mappings from a topological space X to a proximity space Y converges proximally to f, then f is continuous.*

We next consider continuous convergence. A subnet $(x_{n_i} : i \in E)$ of a net $(x_n : n \in D)$ is said to be *isotone* iff the mapping $i \to n_i$ of E into D is order preserving.

(16.7) DEFINITION. *Let X and Y be topological spaces. A net $(f_n: n \in D)$ of mappings from X to Y converges continuously to $f \in Y^X$ iff for every isotone subnet $(f_{n_i}: i \in E)$ of (f_n) and every net $(x_i: i \in E)$ in X converging to x, $(f_{n_i}(x_i))$ converges to $f(x)$.*

(16.8) THEOREM. *Let X be a compact topological space, (Y, δ) a separated proximity space and $(f_n: n \in D)$ a net of continuous mappings from X to Y. If (f_n) converges continuously to a continuous function $f \in Y^X$, then (f_n) converges proximally to f.*

Proof. Let $\mathscr{Y}$ be the Smirnov compactification of Y and let the closure of subsets of Y be taken in $\mathscr{Y}$. Suppose $A \subset X$, $B \subset Y$ and $f_n(A) \delta B$, where n is an arbitrary member of a cofinal subset H of D. Now $f_n(A) \delta B$ implies $\overline{f_n(A)} \cap \bar{B} \neq \varnothing$. Since f_n is continuous and $\bar{A}$ is compact, $f_n(\bar{A})$ is compact and hence closed ($\mathscr{Y}$ is Hausdorff). Thus $f_n(\bar{A})$ contains $\overline{f_n(A)}$, and so meets $\bar{B}$. Pick $x_n \in \bar{A}$ with $f_n(x_n) = y_n \in \bar{B}$. Now the net $(x_n: n \in H)$ has a subnet $(x_{n_i}: i \in E)$ converging to some $x \in \bar{A}$ and, without loss of generality, we may assume this subnet to be isotone. By continuous convergence, $(f_{n_i}(x_{n_i})) \to f(x)$ and, since $f_{n_i}(x_{n_i}) \in \bar{B}$ for all $i \in E$, $f(x) \in \bar{B}$. Thus $f(\bar{A}) \cap \bar{B} \neq \varnothing$, implying that $\overline{f(A)} \cap \bar{B} \neq \varnothing$. We therefore have $f(A) \delta B$, and (f_n) converges to f proximally.

(16.9) THEOREM. *Let X be a topological space and (Y, δ) a separated proximity space. If a net $(f_n: n \in D)$ of continuous mappings from X to Y converges proximally to $f \in Y^X$, then (f_n) converges continuously to f.*

Proof. Let $\mathscr{Y}$ be the Smirnov compactification of Y, and let the closures of subsets of Y be taken in $\mathscr{Y}$. If the conclusion is not true, then there exists an isotone subnet $(f_{n_i}: i \in E)$ of $(f_n: n \in D)$ and a net $(x_i: i \in E)$ converging to $x \in X$ such that $(f_{n_i}(x_i))$ does not converge to $f(x)$. Because Y is regular, there are neighbourhoods V and W of $f(x)$ such that $\bar{W} \subset V$, and a cofinal subset C of E such that $f_{n_i}(x_i) \notin V$ for all $i \in C$. Since f is continuous by (16.6), there exists an i_o such that $f(x_i) \in \bar{W}$ for all $i \geqslant i_o$. Set $A = \{x_i: i \in C, i \geqslant i_o\}$ and $B = \{f_{n_i}(x_i): i \in C\}$. Then

$$\bar{B} \subset \overline{(\mathscr{Y} - V)} = \mathscr{Y} - V \quad \text{and} \quad \overline{f(A)} \subset \bar{W}.$$

Hence $\overline{f(A)} \cap \bar{B} = \varnothing$, so that $f(A) \delta B$. However, for each $i \in C$,

$f_{n_i}(A) \cap B \neq \varnothing$; that is, it is false that eventually $f_n(A) \, \delta \, B$. Thus (f_n) does not converge to f proximally.

17. Unified theories of topology, proximity and uniformity

E. H. Moore, in his New Haven Colloquium lectures in 1906, said 'the existence of analogies between the central features of various theories implies the existence of a general theory which underlies the particular theories and unifies them with respect to those central features'. Since several features are common to topology, proximity and uniformity, one would expect the existence of general structures which include these as special cases. In this section we briefly outline, without proof, two such general structures.

In one such general theory, the idea of a *transitive relation* between subsets of a set X is taken as the basis. If (X, τ) is a topological space, one can define a transitive order between the subsets of X as follows:

(17.1) $\qquad A < B \quad \text{iff} \quad A \subset \text{Int}\,(B).$

It is possible to give precise axioms for this order which will determine the topology in a unique way, namely

(17.2) $\qquad G \in \tau \quad \text{iff} \quad G < G.$

One such set of axioms is the following:

(i) $\varnothing < \varnothing$ and $X < X$.
(ii) $A < B$ and $A' < B'$ implies $A \cap A' < B \cap B'$.
(iii) $A_i < B_i (i \in I)$ implies $\bigcup_{i \in I} A_i < \bigcup_{i \in I} B_i$ for any index set I.

For the case in which (X, δ) is a proximity space, we have already seen in Section 3 that the transitive relation $\ll$ can be used to provide an alternate description of proximity.

Finally, if $(X, \mathscr{U})$ is a uniform space, then Theorem (10.5) suggests a transitive order defined in the following manner:

(17.3) $\qquad A < B \quad \text{iff} \quad U[A] \subset B \quad \text{for some} \quad U \in \mathscr{U}.$

In order to find the axioms for a general structure encompassing topology, proximity and uniformity, one has merely to note the significant common properties satisfied by (17.1), $\ll$ and (17.3). The following definition arose from such considerations:

(17.4) DEFINITION. *A* topogenous order *on a set X is a relation on the power set of X, denoted by* $<$, *satisfying*:

  (i) $\varnothing < \varnothing$ *and* $X < X$.
  (ii) $A < B$ *implies* $A \subset B$.
  (iii) $A \subset A' < B' \subset B$ *implies* $A < B$.
  (iv) $A < B$ *and* $A' < B'$ *together imply* $A \cap A' < B \cap B'$ *and*

$$A \cup A' < B \cup B'.$$

(17.5) DEFINITION. *A* syntopogenous structure *on a set X is a family* $\mathscr{S}$ *of topogenous orders on X satisfying*:

  (S_1) *If* $<_1$ *and* $<_2$ *belong to* $\mathscr{S}$, *there exists a* $<$ *in* $\mathscr{S}$ *such that* $A < B$ *whenever* $A <_1 B$ *or* $A <_2 B$.

  (S_2) *If* $<$ *belongs to* $\mathscr{S}$, *there exists a* $<'$ *in* $\mathscr{S}$ *such that* $A < B$ *implies the existence of a C satisfying* $A <' C <' B$.

With certain particular classes of syntopogenous structures, it is possible to assign topologies, proximities and uniformities. We shall not, however, develop this theory, but instead refer the interested reader to the literature.

Another approach to the unification of these three concepts is the 'neighbourhood' approach, analogous to that of Hausdorff. Let X be any set, $\mathscr{P}$ its power set and $\mathscr{T} = \mathscr{T}(X)$ the set of all singleton subsets of X. Consider a collection $\mathscr{M}$ of subsets of X such that $\mathscr{T} \subset \mathscr{M} \subset \mathscr{P}$. For $U, V \in \mathscr{P}^{\mathscr{M}}$, let $U \cap V$ denote that member of $\mathscr{P}^{\mathscr{M}}$ which satisfies

$$(U \cap V)[A] = U[A] \cap V[A] \quad \text{for each} \quad A \in \mathscr{M}.$$

(17.6) DEFINITION. $\Sigma \subset \mathscr{P}^{\mathscr{M}}$ *is a* GT-structure (*GT standing for 'generalized topological'*) *on X with respect to* $\mathscr{M}$ *iff the following conditions are satisfied*:

  (i) *If* $U \in \Sigma$, *then* $A \subset U[A]$ *for each* $A \in \mathscr{M}$.
  (ii) *If* $U, V \in \Sigma$, *then* $(U \cap V) \in \Sigma$.
  (iii) *If* $U \in \Sigma$, *then for each* $A \in \mathscr{M}$ *there is a* $V \in \Sigma$ (*depending on A*) *such that* $B \in \mathscr{M}$ *and* $B \subset V[A]$ *together imply the existence of a* $W \in \Sigma$ (*depending on B*) *for which* $W[B] \subset U[A]$.

(iv) *If $U \in \mathscr{P}^{\mathscr{M}}$ and, for every $A \in \mathscr{M}$, there is a $V \in \Sigma$ (depending on A) such that $V[A] \subset U[A]$, then $U \in \Sigma$.*

The pair (X, Σ) is called a *GT-space*.

With every GT-space (X, Σ) is associated a 'proximity' relation δ_Σ in the following manner:

(17.7) DEFINITION. *If $A \in \mathscr{M}$ and $B \in \mathscr{P}$, define*

$$A \, \delta_\Sigma B \quad iff \quad U[A] \cap B \neq \varnothing \quad for \; every \quad U \in \Sigma.$$

Σ is said to be *symmetric* iff $A \, \delta_\Sigma B$ implies $B \, \delta_\Sigma A$ whenever $A, B \in \mathscr{M}$.

It is easy to see that a *GT*-structure is a generalization of a topological structure; for if $\mathscr{M} = \mathscr{T}$, then 17.6(i)–(iv) are precisely the usual neighbourhood axioms of Hausdorff, where the fundamental neighbourhood system at x is $\{U[x]: U \in \Sigma\}$. It can also be verified that for $A \subset X$,

(17.8) $$\bar{A} = \{x : x \, \delta_\Sigma A\}$$

defines a Kuratowski closure operator. Thus every *GT*-structure on X induces a topology on X. Conversely, given a topological space (X, τ), we may define a *GT*-structure Σ on X by

$$\Sigma = \{U \in \mathscr{P}^{\mathscr{T}(X)} : x \in \mathrm{Int}\,(U[x]) \text{ for each } x \in X\}.$$

Let us now consider a symmetric *GT*-space (X, Σ) with respect to $\mathscr{M} = \mathscr{P}$. In this case, conditions 17.6(i)–(iv) and the symmetry condition (17.7) are equivalent to the following:

(17.9) (i) $U \in \Sigma$ implies $A \subset U[A]$ for each $A \subset X$.

(ii) $U, V \in \Sigma$ implies $(U \cap V) \in \Sigma$.

(iii) $U \in \Sigma$ implies that for each $A \subset X$, there is a $V \in \Sigma$ (depending on A) such that $V[V[A]] \subset U[A]$.

(iv) If $U \in \mathscr{P}^{\mathscr{P}}$ and, for each $A \subset X$, there exists a $V \in \Sigma$ (depending on A) such that $V[A] \subset U[A]$, then $U \in \Sigma$.

(v) $U \in \Sigma$, $A \subset X$, $B \subset X$ and $U[A] \cap B = \varnothing$ together imply the existence of a $V \in \Sigma$ (depending on A and B) such that $V[B] \cap A = \varnothing$.

It can be verified that in this case, δ_Σ is a proximity on X. Conversely, given a proximity space (X, δ), the set Σ of all functions $U \in \mathscr{P}^{\mathscr{P}}$ such that $A \, \delta \, (X - U[A])$ for every $A \subset X$ yields a symmetric GT-structure on X.

Finally, we consider uniform structures. For $U \in \mathscr{P}^X$, define $U \in \mathscr{P}^{\mathscr{P}}$ by

$$(17.10) \qquad\qquad U[A] = \bigcup_{x \in A} U[x].$$

If, in (17.6), we set $\mathscr{M} = \mathscr{T}$ and require that the mappings involved be independent of x, y etc., we obtain the following set of conditions:

(17.11)   (i)   If $U \in \Sigma$, then $x \in U[x]$ for each $x \in X$.

       (ii)   If $U, V \in \Sigma$, then $(U \cap V) \in \Sigma$.

       (iii)   If $U \in \Sigma$, then there exists a $V \in \Sigma$ such that $V[V[x]] \subset U[x]$ for each $x \in X$.

       (iv)   If, for $U \in \mathscr{P}^X$, there exists a $V \in \Sigma$ such that $V[x] \subset U[x]$ for all $x \in X$, then $U \in \Sigma$.

In this case, Σ is a quasi-uniform base. If in addition Σ is symmetric i.e.

       (v) $U \in \Sigma$ assures the existence of a $V \in \Sigma$ such that $y \in V[x]$ implies $x \in U[y]$,

then Σ is a uniformity base.

Conversely, given a uniform structure $\mathscr{U}$ on X, one can define a symmetric GT-structure Σ on X in the following manner: for $U' \in \mathscr{U}$ and $x \in X$, define $U \in \mathscr{P}^X$ by

$$U[x] = \{y \in X : (x, y) \in U'\}.$$

Definition (17.10) then supplies the desired element of $\mathscr{P}^{\mathscr{P}}$, with Σ consisting of all such elements.

18. Sequential proximity

In this section, we briefly consider nearness relations that can be introduced in Fréchet spaces, i.e. spaces in which convergence of sequences is axiomatized. We motivate the discussion by considering sequences in a separated proximity space (X, δ). If (x_n) and (y_n) are sequences of elements in X, define

(18.1) $(x_n) \, \mathbf{n} \, (y_n)$ iff $\{x_{k_n}\} \, \delta \, \{y_{k_n}\}$ for every increasing sequence (k_n) of natural numbers.

Using Greek letters ξ, η etc. to denote sequences in X, the nearness relation $\mathbf{n}$ of (18.1) can easily be shown to satisfy the following properties, the first three of which imply that $\mathbf{n}$ is an equivalence relation:

(18.2) (i) $\xi \mathbf{n} \xi$ for all ξ.

(ii) $\xi \mathbf{n} \eta$ implies $\eta \mathbf{n} \xi$.

(iii) $\xi \mathbf{n} \xi'$ and $\xi' \mathbf{n} \xi''$ together imply $\xi \mathbf{n} \xi''$.

(iv) $(x) \mathbf{n} (y)$ iff $x = y$.

(v) $(x_n) \mathbf{n} (y_n)$ implies $(x_{k_n}) \mathbf{n} (y_{k_n})$ for every subsequence (k_n) of N.

(vi) If $(x_n) \mathbf{\not{n}} (y_n)$, then there exists an increasing sequence (k_n) in N such that for every subsequence

$$(k_{l_n}) \quad \text{of} \quad (k_n), \quad (x_{k_{l_n}}) \mathbf{\not{n}} (y_{k_{l_n}}).$$

(18.3) THEOREM. *If (X, d) is a metric space and δ is the induced metric proximity on X, then*

$$(x_n) \mathbf{n} (y_n) \quad \text{iff} \quad \lim_{n \to \infty} d(x_n, y_n) = 0.$$

Proof. Clearly, if $\lim\limits_{n \to \infty} d(x_n, y_n) = 0$ then $(x_n) \mathbf{n} (y_n)$. Conversely, suppose $\lim\limits_{n \to \infty} d(x_n, y_n) \neq 0$. Without loss of generality, we may suppose that $d(x_n, y_n) \geqslant \epsilon > 0$ for all $n \in N$. Then one of two cases can arise, which we shall consider separately:

(i) Suppose there is a finite set $\{x_1, \ldots, x_k\}$ such that

$$\min_{1 \leqslant i \leqslant k} \{d(x_n, x_i)\} < \epsilon/4 \quad \text{for} \quad n \in N,$$

or similarly a finite set of y's. Then there exists an x_{i_o} and an increasing sequence of natural numbers (p_n) such that

$$d(x_{p_n}, x_{i_o}) < \epsilon/4.$$

But $d(y_{p_m}, x_{p_n}) \geqslant d(y_{p_m}, x_{p_m}) - d(x_{p_m}, x_{p_n}) > \epsilon - \epsilon/2 = \epsilon/2$, so that $(x_{p_n}) \mathbf{\not{n}} (y_{p_n})$.

(ii) Suppose there exists an increasing sequence (r_n) of natural numbers such that $d(x_{r_n}, x_{r_m}) \geqslant \epsilon/4$ and $d(y_{r_n}, y_{r_m}) \geqslant \epsilon/4$ for n, $m \in N$, $n \neq m$. Let $k_1 = 1$. Define k_{n_o+1} inductively such that $d(x_{k_i}, y_{k_j}) \geqslant \epsilon/8$ for $i \neq j$, where $i, j \geqslant n_o$. There exist at most finitely many $x \in \{x_n : n \geqslant k_{n_o}\}$ such that

$$d(x_n, \{\{x_{k_i}\} \cup \{y_{k_i}\} : i \leqslant n_o\}) < \epsilon/8,$$

and likewise for $\{y_n \colon n \geqslant k_{n_0}\}$. Thus there is an $n' > k_{n_0}$ such that the distances between each of $x_{n'}$, $y_{n'}$ and the set $\{x_{k_i}\} \cup \{y_{k_i}\}$, where $i \leqslant n_0$, are greater than $\epsilon/8$. Set $k_{n_0+1} = n'$. Then, for the sequence (k_n) in N, defined inductively in this way, $(x_{k_n}) \not\!\mathbf{n} \, (y_{k_n})$.

(18.4) DEFINITION. *A sequence (x_n) in a separated proximity space (X, δ) is Cauchy iff $(x_n) \, \mathbf{n} \, (x_{k_n})$ for every increasing sequence (k_n) in N.*

Every convergent sequence is Cauchy, although the converse is not true. If every Cauchy sequence in a metric space converges, then the metric space is complete in the usual sense.

(18.5) THEOREM. *Let X be a complete metric space and let $\mathscr{X}$ be its Smirnov compactification. Then the first axiom of countability C_I is not satisfied at points of $\mathscr{X} - X$.*

Proof. If C_I is satisfied at $x \in \mathscr{X} - X$, there exists a sequence (x_n) in X converging to x. But (x_n) is Cauchy and, since X is complete, we have that $x \in X$, a contradiction.

Having seen the usefulness of the concept of nearness of sequences in a proximity space, we now turn our attention to the axiomatization of the nearness relation. Properties in (18.2) suggest the following definition:

(18.6) DEFINITION. *A nearness relation $\mathbf{n}$, between sequences of elements of a set X, which satisfies 18.2(i)–(v) is called a $\mathscr{UL}$-structure. The pair $(X, \mathbf{n})$ is called a $\mathscr{UL}$-space. If in addition $\mathbf{n}$ satisfies 18.2(vi), then $(X, \mathbf{n})$ is a $\mathscr{UL}^*$-space.*

Theorem (18.3), a special case of (18.1), gives an example of a $\mathscr{UL}^*$-space. We now present two further examples:

(18.7) EXAMPLES. (i) A trivial $\mathscr{UL}^*$-structure can be defined on X by letting

$$(x_n) \, \mathbf{n} \, (x_n') \quad \text{iff} \quad x_n = x_n' \text{ eventually.}$$

(ii) Let $(Y, \mathbf{n})$ be a $\mathscr{UL}$-(resp. $\mathscr{UL}^*$-) space and let $\mathscr{F}$ be a collection of functions from a set X to Y which separates points; that is, $x_1 \neq x_2$ for $x_1, x_2 \in X$ implies the existence of an $f \in \mathscr{F}$ such that $f(x_1) \neq f(x_2)$. Define a nearness relation $\mathbf{N}$ on X by

$$(x_n) \, \mathbf{N} \, (x_n') \text{ iff } (f(x_n)) \, \mathbf{n} \, (f(x_n')) \quad \text{for all} \quad f \in F.$$

Then $(X, \mathbf{N})$ is a $\mathscr{UL}$-(resp. $\mathscr{UL}^*$-) space.

(18.8) DEFINITION. *A sequence (x_n) in a $\mathscr{UL}$-space $(X, \mathbf{n})$ converges to $x \in X$ iff $(x_n)\,\mathbf{n}\,(x)$.*

(18.9) DEFINITION. *A* Fréchet *or $\mathscr{L}$-space is a pair $(X, \mathscr{L})$ where X is any set and $\mathscr{L}$ is a collection of sequences, converging to certain points called limits, satisfying the following conditions:*

(i) *If $x_n = x$ for all $n \in N$, then $(x_n) \in \mathscr{L}$ and $\lim x_n = x$.*

(ii) *$\lim x_n = x$ implies $(x_{n_i}) \in \mathscr{L}$ and $\lim x_{n_i} = x$, for every subsequence (x_{n_i}) of (x_n).*

$(X, \mathscr{L})$ is called an $\mathscr{L}^*$-space iff whenever $(x_n) \in \mathscr{L}$ does not converge to x, there exists a subsequence $(x_{n_i}) \in \mathscr{L}$ such that no subsequence of it converges to x.

The following result is obvious.

(18.10) THEOREM. *A $\mathscr{UL}$-($\mathscr{UL}^*$-) space with the convergence relation of (18.8) defined on it is an $\mathscr{L}$-($\mathscr{L}^*$-) space, where $\mathscr{L}$ is the set of all convergent sequences.*

We now turn to the problem of defining a nearness relation in an $\mathscr{L}$-($\mathscr{L}^*$-) space.

(18.11) DEFINITION. *A nearness relation $\mathbf{n}$ in an $\mathscr{L}$-($\mathscr{L}^*$-) space is* compatible *iff the convergence induced by $\mathbf{n}$ coincides with the original.*

There may exist several compatible $\mathscr{UL}$-($\mathscr{UL}^*$-) structures on a given $\mathscr{L}$-($\mathscr{L}^*$-) space. A partial order can be defined on these structures, but we shall not go into the details. Suffice it to say that given an $\mathscr{L}$-($\mathscr{L}^*$-) space X, we may introduce two extreme compatible $\mathscr{UL}$-structures:

(18.12) (a) $(x_n)\,\mathbf{n}_o\,(x_n')$ iff both sequences converge to the same limit.

(b) $(x_n)\,\mathbf{n}_1\,(x_n')$ iff for each increasing sequence (i_n) in N, either (x_{i_n}) and (x_{i_n}') are both divergent or both converge to the same limit.

Given a $\mathscr{UL}$-space $(X, \mathbf{n})$, we may define

(18.13) $(x_n)\,\mathbf{n}^*\,(y_n)$ iff each increasing sequence (i_n) in N contains a subsequence (j_n) such that $(x_{j_n})\,\mathbf{n}\,(y_{j_n})$.

Then $(X, \mathbf{n}^*)$ is a $\mathscr{UL}^*$-space. If $(X, \mathscr{L}^*)$ is compact (i.e. every sequence in X has a convergent subsequence), then $\mathbf{n}_o^* = \mathbf{n}_1^*$; that is, it has a unique compatible $\mathscr{UL}^*$-structure.

Finally, we consider $\mathscr{L}$-uniform convergence.

(18.14) DEFINITION. *Given a $\mathscr{UL}$-(or $\mathscr{UL}^*$-) space $(Y, \mathbf{n})$, a sequence (f_n) in Y^X converges $\mathscr{L}$-uniformly to $f \in Y^X$ iff*

$$(f_n(x_n))\, \mathbf{n}\, (f(x_n))$$

for every sequence (x_n) in X.

(18.15) THEOREM. *Let X be a topological space and $(Y, \mathbf{n})$ a $\mathscr{UL}$-space, where $\mathbf{n}$ is induced by the proximity δ on Y. If $f_n \colon X \to Y$ is sequentially continuous at $x_o \in X$ for each $n \in N$ and (f_n) converges $\mathscr{L}$-uniformly to $f \in Y^X$, then f is sequentially continuous at x_o.*

Proof. Suppose (x_n) converges to x_o, but $(f(x_n))$ does not converge to $f(x_o)$. We may then suppose, without loss of generality, that there is a neighbourhood U of $f(x_o)$ such that $f(x_n) \notin U$ for every $n \in N$. Now there exists a neighbourhood U_1 of $f(x_0)$ such that $U_1 \ll U$. Since $(f_n(x_o))$ converges to $f(x_o)$ and f_n is sequentially continuous at x_0, for each $n \in N$ there exists a subsequence (k_n) of N such that $f_n(x_{k_n}) \in U_1$. Thus $(f_n(x_{k_n}))\,\not{\mathbf{n}}\,(f(x_{k_n}))$, contrary to the hypothesis that (f_n) converges uniformly to f.

19. Generalized proximities

Several generalized forms of proximity structures are known in the literature, some of which were introduced even before the appearance of Efremovič proximity spaces. In this section we briefly study some of these generalized proximities and their interrelationships. One of these is studied in some detail in the next section.

Since a compatible proximity can be introduced only in completely regular topological spaces, one wonders whether the axioms of proximity can be relaxed so as to embrace more general topological spaces. Indeed, generalized proximities can in fact be introduced in any topological space, as we shall see presently.

(19.1) DEFINITION. *A binary relation α defined on the power set of X is called a* Leader *or* LE-proximity *on X iff it satisfies the following conditions*:

   (i) $A \alpha (B \cup C)$ *iff* $A \alpha B$ *or* $A \alpha C$, *and*

$$(A \cup B) \alpha C \text{ iff } A \alpha C \text{ or } B \alpha C.$$

   (ii) $A \alpha B$ *implies* $A \neq \varnothing$, $B \neq \varnothing$.

   (iii) $A \alpha B$ *and* $b \alpha C$ *for each* $b \in B$ *together imply* $A \alpha C$.

   (iv) $A \cap B \neq \varnothing$ *implies* $A \alpha B$.

If in addition α satisfies

   (v) $A \alpha B$ *iff* $B \alpha A$,

then α is called a Lodato *or* LO-proximity. *The pair (X, δ), where δ is a LO-proximity, is referred to as a* Lodato space.

(19.2) DEFINITION. *A binary relation β defined on the power set of X is called a* Pervin *or* P-proximity *on X iff β satisfies* 19.1 (i), (ii), (iv) *and*

   (iii′) $A \beta B$ *implies there exists an* $E \subset X$ *such that* $A \beta E$ *and* $(X - E) \beta B$.

It should be noted that symmetry is demanded neither for a LE-proximity nor a P-proximity, and thus care must be taken in writing proofs involving either of these.

Throughout this section we shall use the letter ξ to denote an arbitrary generalized proximity on X. ξ is said to be separated *iff it satisfies the additional condition*

$$x \xi y \quad \text{implies} \quad x = y.$$

A partial order may be defined on the collection of all generalized proximities on a set X by (2.16), namely

$$\xi_1 < \xi_2 \quad \text{iff} \quad A \xi_2 B \quad \text{implies} \quad A \xi_1 B.$$

We now compare α with β. First note that 2.1 (i), namely

$$A \xi B, \ A \subset C \quad \text{and} \quad B \subset D \text{ together imply } C \xi D,$$

holds for $\xi = \alpha$ or β.

(19.3) THEOREM. *Every* P-*proximity β on X is also a* LE-*proximity on X.*

Proof. It is sufficient to show that β satisfies 19.1 (iii). Suppose $A \beta B$ and $b \beta C$ for every $b \in B$. If $A \not\beta C$, then by 19.2 (iii′) there exists an $E \subset X$ such that $A \not\beta E$ and $(X - E) \not\beta C$. Now

$A \beta E$ and $A \beta B$ together imply $B \not\subset E$, i.e. $B \cap (X - E) \neq \varnothing$. If $b \in B \cap (X - E)$, then $b \beta C$ implies $(X - E) \beta C$, a contradiction.

Given $\xi = \alpha$ or β, define

(19.4) $$A^\xi = \{x \in X : x \xi A\}.$$

As in (2.7), one can prove that if $\xi = \alpha$ or β, then (19.4) defines a Kuratowski closure operator on X. In the case that $\xi = \alpha$, the result $(A^\xi)^\xi = A^\xi$ follows from 19.1 (iii). Actually a weaker axiom, namely

(19.5) for $x \in X$, $x \alpha B$ and $b \alpha C$ for all $b \in B$ implies $x \alpha C$,

is sufficient to guarantee the idempotence of the operator ξ. Thus either $\xi = \alpha$ or β yields a topology $\tau = \tau(\xi)$, and we say that τ and ξ are compatible. Moreover, it is easy to show that

(19.6) $$A \xi B \quad \text{iff} \quad A \xi B^\xi.$$

In contrast to (Efremovič) proximity structures, which are compatible with completely regular spaces, we have the following result:

(19.7) THEOREM. *Every topological space (X, τ) has a compatible LE- or P-proximity ξ_0 given by*

(19.8) $$A \xi_0 B \quad \text{iff} \quad A \cap \bar{B} \neq \varnothing.$$

Moreover, ξ_0 is the largest compatible LE- or P-proximity.

Proof. In view of Theorem (19.3) we need only verify 19.2 (iii'), since the other axioms follow readily. If $A \not\xi_0 B$, then $A \cap \bar{B} = \varnothing$. Set $E = \bar{B}$. Then $A \cap \bar{E} = \varnothing$ and $(X - E) \cap \bar{B} = \varnothing$, i.e. $A \not\xi_0 E$ and $(X - E) \not\xi_0 B$. That $\tau = \tau(\xi_0)$ follows from the fact that $x \xi_0 A$ iff $x \in \bar{A}$. Finally, if ξ is any LE- or P-proximity, then from (19.6) we have that $A \xi B$ implies $A \cap \bar{B} = \varnothing$, i.e. $A \not\xi_0 B$.

A topological space (X, τ) is R_o iff either of the following equivalent conditions is satisfied:

(19.9) (i) $x \in \bar{y}$ iff $y \in \bar{x}$.

  (ii) $x \in G \in \tau$ implies $\bar{x} \subset G$.

(19.10) THEOREM. *If α is any LO-proximity, then $\tau(\alpha)$ is necessarily R_o. Conversely, a compatible LO-proximity α_1 can be defined on every R_o-space by*

(19.11) $$A \alpha_1 B \quad \text{iff} \quad \bar{A} \cap \bar{B} \neq \varnothing.$$

Furthermore, α_1 is the largest compatible LO-proximity.

Proof. That $\tau(\alpha)$ is R_o follows from the fact that $x \in \bar{y}$ iff $x \alpha y$ iff $y \alpha x$ iff $y \in \bar{x}$. To prove that α_1 is a LO-proximity, it suffices to verify 19.1(iii). Suppose $A \alpha_1 B$ and $b \alpha_1 C$ for each $b \in B$. Then $\bar{A} \cap \bar{B} \neq \varnothing$ and $\bar{b} \cap \bar{C} \neq \varnothing$ for each $b \in B$, i.e. there exists a $c \in \bar{C}$ such that $c \in \bar{b}$. Since X is R_o, $b \in \bar{c} \subset \bar{C}$ and hence $\bar{A} \cap \bar{C} \neq \varnothing$, showing that $A \alpha_1 C$. Since $x \in A^{\alpha_1}$ iff $\bar{x} \cap \bar{A} \neq \varnothing$ iff $x \in \bar{A}$, it follows that $\tau = \tau(\alpha_1)$. For every LO-proximity α, $A \alpha B$ iff $\bar{A} \alpha \bar{B}$, and thus α_1 is the largest compatible LO-proximity.

(19.12) COROLLARY. *There exists a LO-proximity which is not a P-proximity.*

Proof. There exist R_o-spaces which are not regular and (19.11) shows that if α_1 were a P-proximity, then it would also be an Efremovič proximity, which is impossible.

A *quasi-uniformity* $\mathcal{Q}$ on a set X is a family of subsets of $X \times X$ which satisfies all the conditions for a uniformity with the possible exception of the symmetry axiom.

The topology $\tau(\mathcal{Q})$ induced by $\mathcal{Q}$ is the family $\{G \subset X : \text{for each} \ x \in G, \text{there exists a} \ U \in \mathcal{Q} \text{ such that } U[x] \subset G\}$. It is known that every topological space (X, τ) has a compatible *Pervin quasi-uniformity* $\mathcal{P}$ consisting of all sets which contain finite intersections of sets of the form:

(19.13) $\qquad S_G = [G \times G] \cup [(X - G) \times X] \quad$ where $\quad G \in \tau$.

The proof of the following theorem is similar to that of Theorem (10.2).

(19.14) THEOREM. *Every quasi-uniformity $\mathcal{Q}$ on X induces a P- (or LE-) proximity ξ on X defined by*

(19.15) $\qquad A \xi B \quad$ iff $\quad (A \times B) \cap U \neq \varnothing \quad$ for every $\quad U \in \mathcal{Q}$.

Moreover, $\tau(\xi) = \tau(\mathcal{Q})$. If $\mathcal{Q} = \mathcal{P}$ (the Pervin quasi-uniformity), then $\xi = \xi_0$, as defined by (19.8).

Let X be a T_1-topological space and A, B be non-empty subsets of X. If

(19.16) $\qquad\qquad (A \cap \bar{B}) \cup (\bar{A} \cup B) = \varnothing$, then $(A \cup B)$ is separated, i.e. not connected in the Hausdorff–Lennes sense. If we write $A \not\gamma B$ to denote this separation, then γ obviously satisfies the axioms given in the following definition:

(19.17) DEFINITION. *A binary relation γ defined on the power set of X is called a* Separation *or* S-proximity *iff the following conditions are satisfied*:

  (i) $A \gamma B$ *implies* $B \gamma A$.
 (ii) $(A \cup B) \gamma C$ *iff* $A \gamma C$ *or* $B \gamma C$.
(iii) $A \gamma B$ *implies* $A \neq \varnothing$, $B \neq \varnothing$.
 (iv) $x \gamma B$ *and* $b \gamma C$ *for every* $b \in B$ *together imply* $x \gamma C$.
  (v) $A \cap B \neq \varnothing$ *implies* $A \gamma B$.
 (vi) $x \gamma y$ *implies* $x = y$.

Clearly 19.1(iii) is stronger than 19.17(iv), and the following result is immediate:

(19.18) THEOREM. *Every separated* LO-*proximity on X is also an* S-*proximity on X.*

We now give an example of an S-proximity which is not a LO-proximity:

(19.19) EXAMPLE. If on the set of real numbers with the usual topology we define γ by (19.16), then γ is strictly larger than α_1 as defined by (19.11). But α_1 is the largest compatible LO-proximity, as shown in Theorem (19.10), so that the S-proximity γ cannot be a LO-proximity.

20. More on Lodato spaces

In the theory of (Efremovič) proximity spaces, one of the most important results (Theorem (7.7)) is the following: every separated proximity space is δ-homeomorphic to a dense subspace of a compact Hausdorff space in which the unique compatible proximity δ_o is given by

$$A \delta_o B \quad \text{iff} \quad \bar{A} \cap \bar{B} \neq \varnothing.$$

A natural question arises which, in fact, provides a motivation for Lodato spaces: does there exist a set of axioms for a binary relation δ on the power set of X such that δ satisfies these axioms iff there is a topological space Y in which X can be topo-logically embedded so that $A \delta B$ in X iff $\bar{A} \cap \bar{B} \neq \varnothing$ in Y? Granting that such an embedding exists, one can easily show that δ must necessarily satisfy the axioms of a LO-proximity as given in Definition (19.1). The converse question was first handled in the literature using a technique involving clusters which is similar to the construction of the Smirnov compactification (Section 7), the difference being that instead of defining a proximity on the set of all clusters, one merely defines a topology via a Kuratowski closure operator. This approach provided an affirmative solution under the restriction that Y be a T_1-space and X be regularly dense in Y. The latter condition was subsequently relaxed, when the notion of a bunch was introduced. Recently the theory of Lodato spaces has been extended with the introduction of 'symmetric generalized uniform structures', which we shall call M-uniform structures. In this section, we outline the theory of Lodato spaces only briefly and refer the interested reader to the literature for a more detailed account.

(20.1) DEFINITION. *A subset X of a topological space (Y, τ) is regularly dense in Y iff given $U \in \tau$ and $p \in U$, there exists a subset A of X with $p \in \bar{A} \subset U$.*

Use of the term 'regularly dense' is justified by the following readily-verified facts:

(i) A regularly dense set is dense.

(ii) If Y is regular, then dense and regularly dense sets co-incide.

Definition (5.4) is used to define a cluster in a Lodato space exactly as it is in a proximity space. One can prove without difficulty that Remarks (5.5) and Lemma (5.6) remain valid in the generalized setting. We are now in a position to present the first solution to the question posed in the opening paragraph.

(20.2) THEOREM. *If δ is a binary relation on the power set of X, then the following statements are equivalent :*

(a) There exists a T_1-space Y in which X can be topologically embedded as a regularly dense subset such that

(20.3) $A \delta B$ in X iff $\bar{A} \cap \bar{B} \neq \varnothing$ in Y.

(b) δ is a separated LO-proximity satisfying the additional axiom:

(20.4) $A \delta B$ implies the existence of a cluster to which both A and B belong.

Proof. The symbol $^-$ will be used to denote closure in Y. In showing that (a) implies (b), we first note that by Theorem (19.10), δ is a LO-proximity on Y, and so induces one on the subspace X. Furthermore, the T_1-axiom is equivalent to the condition that $x \delta y$ implies $x = y$. To prove that (20.4) is satisfied, suppose $A \delta B$ in X, and hence $\bar{A} \cap \bar{B} \neq \varnothing$ in Y. Let $y \in \bar{A} \cap \bar{B}$ and set $\sigma = \{E \subset X : y \in \bar{E}\}$. That σ is a cluster in X will be clear if we can show that $C \delta E$ for every $E \in \sigma$ implies $C \in \sigma$. Suppose on the contrary that $C \notin \sigma$, so that $y \in Y - \bar{C}$. Since X is regularly dense in Y, there exists an $F \subset X$ such that $y \in \bar{F} \subset Y - \bar{C}$. Then $C \bar{\delta} F$ where $F \in \sigma$, a contradiction. Thus σ is a cluster, to which both A and B belong.

In proving the converse, we use the notation of Section 7: $f(x) = \sigma_x$ (the point cluster), $Y = \mathscr{X} = \{\sigma : \sigma$ is a cluster in $X\}$, and $\bar{\mathscr{A}} = \{\sigma \in \mathscr{X} : A \in \sigma\}$ for each $A \subset X$. Clearly f is one-to-one and $f(A) \subset \mathscr{A}$. Define an operator Cl on $\mathscr{X}$ by

(20.5) $\mathrm{Cl}(\mathscr{P}) = \{\sigma \in \mathscr{X} : E \subset X$ absorbing $\mathscr{P}$ implies $E \in \sigma\}$.

Observing that

(20.6) $\mathrm{Cl}(f(A)) = \bar{\mathscr{A}}$ for $A \subset X$,

it becomes a routine verification to show that Cl is a Kuratowski closure operator on $\mathscr{X}$. If $\sigma' \in \mathrm{Cl}(\sigma)$, then clearly $\sigma \subset \sigma'$ and, since clusters are maximal (Lemma (5.6)), $\sigma = \sigma'$. Consequently $\mathscr{X}$ is T_1.

We next prove (20.3). If $A \delta B$, then by (20.4) there exists a $\sigma \in \mathscr{X}$ such that $A, B \in \sigma$. Hence $\sigma \in \mathrm{Cl}(f(A)) \cap \mathrm{Cl}(f(B))$. On the other hand, if $\sigma \in \mathrm{Cl}(f(A)) \cap \mathrm{Cl}(f(B))$, then $A, B \in \sigma$ and so $A \delta B$.

To see that $f(X)$ is regularly dense in $\mathscr{X}$, suppose $\mathscr{P}$ is an open subset of $\mathscr{X}$ and $\sigma \in \mathscr{P}$. Then $\sigma \notin (\mathscr{X} - \mathscr{P}) = \mathrm{Cl}(\mathscr{X} - \mathscr{P})$. By (20.6), there exists an $E \subset X$ such that E belongs to every cluster in $\mathscr{X} - \mathscr{P}$, but $E \notin \sigma$. Hence there exists a $C \in \sigma$ such that $E \bar{\delta} C$. Clearly $\sigma \in \bar{\mathscr{C}}$ and, since E belongs to every cluster in $\mathscr{X} - \mathscr{P}$ and $E \bar{\delta} C$, C belongs to no cluster in $\mathscr{X} - \mathscr{P}$. Thus $\bar{\mathscr{C}} \subset \mathscr{P}$ and we have proven that $f(X)$ is regularly dense in $\mathscr{X}$.

Finally, to show that X is topologically embedded in $\mathscr{X}$, we need only show that f is bicontinuous; that is,

$$x \in A^\delta \quad \text{iff} \quad f(x) \in Cl(f(A)) = \overline{\mathscr{A}} \quad \text{for all} \quad A \subset X.$$

But clearly, $x \in A^\delta$ iff $x\,\delta\,A$ iff $A \in \sigma_x$ iff $f(x) = \sigma_x \in \mathscr{A}$.

(20.7) DEFINITION. *A non-empty collection σ of subsets of a Lodato space (X, δ) is called a* bunch *iff the following conditions are satisfied*:

  (i) *If $A, B \in \sigma$, then $A\,\delta\,B$.*
  (ii) *If $(A \cup B) \in \sigma$, then $A \in \sigma$ or $B \in \sigma$.*
  (iii) *If $A \in \sigma$ and $a\,\delta\,B$ for every $a \in A$, then $B \in \sigma$.*

(20.8) REMARKS. (a) Although every cluster is a bunch, a bunch need not be a cluster. To illustrate the latter statement, take the family of all infinite subsets of an infinite set X with cofinite topology and LO-proximity α_1 (19–11).

(b) $X \in \sigma$ for every bunch σ, as a consequence of 20.7(iii). Although redundant, this condition was originally included in the definition of a bunch.

Consider a family $\mathscr{B}$ of bunches in (X, δ) which satisfies the two conditions:

(20.9) (i) $A\,\delta\,B$ implies there is a $\sigma \in \mathscr{B}$ such that $A, B \in \sigma$.

  (ii) If $\sigma, \sigma' \in \mathscr{B}$ and either $A \in \sigma$ or $B \in \sigma'$ for all subsets A, B of X such that $A \cup B = X$, then $\sigma = \sigma'$.

By introducing a Kuratowski closure operator on $\mathscr{B}$ using (20.5), one can prove the following result in a manner similar to the proof of Theorem (20.2).

(20.10) THEOREM. *Given a set X and a binary relation δ on the power set of X, the following statements are equivalent*:

(a) *There exists a Hausdorff space Y in which X can be topologically embedded so that (20.3) holds.*

(b) *(X, δ) is a separated LO-proximity space possessing a family $\mathscr{B}$ satisfying (20.9).*

In the theory of proximity spaces, interesting relationships exist between uniform structures and proximity structures, as we saw in Chapter 3. The notion of a Mozzochi or M-uniform structure is now introduced, playing a role in Lodato spaces similar to that played by uniform structures in proximity spaces.

(20.11) DEFINITION. *A non-empty subset $\mathscr{U}$ of the power set of $X \times X$ is a* Mozzochi *or* M-uniform structure *on X iff the following axioms are satisfied:*

  (i) $\Delta \subset U$ *for each $U \in \mathscr{U}$.*

  (ii) $U \in \mathscr{U}$ *implies $U = U^{-1}$.*

  (iii) *For every $A \subset X$ and U, $V \in \mathscr{U}$, there exists a $W \in \mathscr{U}$ such that $W[A] \subset U[A] \cap V[A]$.*

  (iv) *For every pair of subsets A, B of X and every $U \in \mathscr{U}$,*

$$V[A] \cap B \neq \varnothing \quad \text{for every} \quad V \in \mathscr{U}$$

implies the existence of an $x \in B$ and $W \in \mathscr{U}$ such that $W[x] \subset U[A]$.

  (v) *$U \in \mathscr{U}$ and $U \subset V = V^{-1} \subset X \times X$ implies $V \in \mathscr{U}$.*

The pair $(X, \mathscr{U})$ is called an M-uniform space. *It is* separated *iff*

  (vi) $\bigcap_{U \in \mathscr{U}} U = \Delta$.

A subfamily $\mathscr{B}$ of an M-uniformity $\mathscr{U}$ is a base *for $\mathscr{U}$ iff each member of $\mathscr{U}$ contains a member of $\mathscr{B}$.*

Clearly, the family of all symmetric entourages of a uniformity (and, in fact, of an A–N uniformity) forms an M-uniform structure. The existence of M-uniformities which are not uniformities will be demonstrated shortly (see 20.19).

The proof of the following theorem, being reasonably straightforward, is omitted.

(20.12) THEOREM. *Every* M-uniform *space $(X, \mathscr{U})$ has an associated topology $\tau = \tau(\mathscr{U})$ defined by*

(20.13) *$G \in \tau$ iff for each $x \in G$, there exists a $U \in \mathscr{U}$ such that*

$$U[x] \subset G.$$

Alternatively, the operator Cl defined by

(20.14) $\mathrm{Cl}\,(A) = \bigcap_{U \in \mathscr{U}} U[A]$

is a Kuratowski closure operator yielding τ. The topology $\tau(\mathscr{U})$ is necessarily R_0; $\tau(\mathscr{U})$ is T_1 if and only if $\mathscr{U}$ is separated.

Just as every uniform structure has an associated proximity relation, it can be shown that every M-uniform structure $\mathscr{U}$ induces a LO-proximity $\delta = \delta(\mathscr{U})$ defined by

(20.15) $A \, \delta \, B$ iff $U[A] \cap B \neq \varnothing$ for every $U \in \mathscr{U}$

iff $(A \times B) \cap U \neq \varnothing$ for every $U \in \mathscr{U}$.

(If $\delta = \delta(\mathscr{U})$, then δ and $\mathscr{U}$ are said to be compatible. It is clear that whenever one is separated, so is the other.) In fact, an even stronger result is valid, the proof of which is similar to that of Theorem (13.14) and is thus omitted.

(20.16) THEOREM. *Let δ be a binary relation on the power set of X and $\mathscr{U}$ be a collection of symmetric subsets of $X \times X$ such that δ and $\mathscr{U}$ satisfy* (20.15). *Then δ is a LO-proximity if and only if $\mathscr{U}$ is a base for an M-uniform structure. Furthermore, $\tau(\delta) = \tau(\mathscr{U})$.*

(20.17) THEOREM. *Every Lodato space (X, δ) has a compatible M-uniform structure $\mathscr{U}$ (i.e. $\delta = \delta(\mathscr{U})$).*

Proof. For every pair of subsets A and B of X, define

$$U_{A,B} = X \times X - [(A \times B) \cup (B \times A)].$$

Set $\mathscr{V} = \{U_{A,B} : A \, \delta \, B\}$. Then each member of $\mathscr{V}$ is clearly symmetric, so that 20.11(ii) is satisfied. Now if $A \, \delta \, B$, then $U_{A,B}[A] \cap B = \varnothing$. Conversely, if $U_{C,D}[A] \cap B = \varnothing$ for some pair C, D such that $C \, \delta \, D$, then either $A \subset C$ and $B \subset D$ or $A \subset D$ and $B \subset C$. In either case, $A \, \delta \, B$. Hence by the previous theorem, $\mathscr{V}$ is a base for an M-uniform structure $\mathscr{U}_0$ and $\delta = \delta(\mathscr{U}_0)$.

(20.18) COROLLARY. *The topology τ on X is the topology induced by some M-uniform structure if and only if τ is R_0.*

Proof. This result follows from Theorems (19.10) and (20.17).

(20.19) REMARK. Since there are R_0-spaces which are not completely regular, the above corollary shows the existence of M-uniformities which are not uniformities.

(20.20) DEFINITION. *Given a Lodato space (X, δ), the collection of all compatible M-uniform structures on X is a LO-proximity class of M-uniformities and is denoted by $\Gamma(\delta)$.*

(20.21) THEOREM. *Let (X, δ) be a Lodato space. Then $\mathcal{U}_0$, as defined in (20.17), is the smallest member of $\Gamma(\delta)$ (under the usual partial order of set inclusion).*

Proof. Suppose $U_{A,B} \in \mathcal{U}_o$ and $\mathcal{V} \in \Gamma(\delta)$. Since $A \not\delta B$, we know from (20.15) that there exists a $V \in \mathcal{V}$ such that $(A \times B) \cap V = \varnothing$. But V is symmetric, so that $(B \times A) \cap V = \varnothing$ also. We therefore have $V \subset U_{A,B}$, showing that $U_{A,B} \in \mathcal{V}$.

In contrast to the situation with a proximity class of uniformities, every LO-proximity class of M-uniform structures possesses a largest element and hence forms a complete lattice. That such is the case is evident from the next theorem.

(20.22) THEOREM. *The union of an arbitrary family of M-uniform structures belonging to the same proximity class $\Gamma(\delta)$ forms a base for a member of $\Gamma(\delta)$.*

Proof. If $\mathscr{B}$ denotes the union of such a family, then $A \not\delta B$ iff $U[A] \cap B \neq \varnothing$ for every $U \in \mathscr{B}$. Consequently, by Theorem (20.16), $\mathscr{B}$ is a base for some M-uniform structure which is a member of $\Gamma(\delta)$.

Notes

16. The concept of proximal convergence and Theorems (16.2), (16.3) and (16.5) are due to Leader [54]. The proof of Theorem (16.3) appearing in this section, however, is that of Njåstad [80]. Leader has also shown that if a net (f_n) is in fact a sequence, then proximal convergence and uniform convergence coincide. Njåstad has generalized this result to the case in which the directed set of the net is linear and the range space possesses a generalized uniform structure. Theorems (16.8) and (16.9) were proved by Wolk [122]. The definition of continuous convergence given here is stronger than that given in Kelley [J] (p. 241, Problem M).

17. As pointed out in the text, we have presented only bare outlines of two general theories. Császár [12] first gave his theory of syntopogenous structures. Subsequently, Doĭčinov [15] used the neighbourhood approach.

18. Mrówka [75] introduced sequential proximity in a proximity space and proved Theorems (18.3), (18.5) and (18.15). An axiomatic treatment of sequential proximity was first given by Goetz [28]. For an up-to-date account of this subject see Goetz [29], where a summary of the work of Poljakov [91] is also included.

19. We have taken the liberty to deviate from the names originally used by authors to describe their generalized proximities, in that we have used the terms LE-proximity and LO-proximity space rather than 'topological d-space' and 'symmetric generalized proximity space' respectively. The definition of LE-proximity was first given by Leader [57], his motivation being an explicit formulation of the proximity relation induced on the product of proximity spaces by the proximity relations on the co-ordinate spaces. Lodato [63, 64, 65] introduced and developed the LO-proximity relation. Mozzochi [72] has since carried out an extensive study of LO-proximity, a brief account of which is given in Section 20. For information and references concerning the R_0-axiom, consult *Ann. Univ. Sci. Budapest. Eötvös Sect. Math.* **10,** 53–4 (1967). Pervin [84] introduced the P-proximity relation, calling it 'quasi-proximity'; Steiner [110] later corrected an error in this paper. The S-proximity relation was conceived independently by Krishna Murti [52], Szymanski [113] and Wallace [116, 117]. For an interesting discussion of some of the generalizations of proximity, see Pervin [85]. Interrelationships between various generalized proximities discussed in this section are due to the authors.

Recently, Mattson [67] has studied S-proximity spaces in relation to the extended topologies of Hammer. Around 1964, Hayashi [32, 33] introduced two new proximities: (i) paraproximity, and (ii) pseudo-proximity. Given a paraproximity he defines a topology by calling a set U open iff $U\,\delta\,(X-U)$, while in the case of pseudo-proximity U is defined to be open iff $x\,\delta\,(X-U)$ for each $x \in U$ (cf. (2.4)). The topology of a paraproximity is necessarily completely normal. Only two axioms are used to define pseudo-proximity:

(i) $A\,\delta\,\varnothing$ for every $A \subset X$, and
(ii) $C\,\delta\,(A \cup B)$ iff $C\,\delta\,A$ or $C\,\delta\,B$.

Recently, Fedorčuk [22] has defined θ-proximity in regular topological spaces.

20. Theorems (20.2) and (20.10) were proved by Lodato [63, 64], with the notion of a bunch being introduced in the second paper. In his dissertation, Mozzochi [72] generalized many results of Chapter 3 to Lodato spaces with the introduction of 'symmetric generalized uniform structures', which we have called M-uniform structures. Only a small part of his work is outlined in this section. LO-spaces have recently been used to derive a rather general theorem concerning the extensions of continuous functions from dense subspaces (see [138]). Example (20.8) was discovered by P. L. Sharma.

GENERAL REFERENCES

A ALEXANDROFF, P. (1939). On bicompact extensions of topological spaces, *Mat. Sb.* **5** (47), 403–423 (In Russian; German summary).
B FREUDENTHAL, H. (1942). Neuaubau der Endentheorie, *Ann. of Math.* **43** (2), 261–279.
C GAAL, S. A. (1964). *Point Set Topology*, Academic Press, New York.
D GÁL, I. S. (1959). Uniformizable spaces with a unique structure, *Pacif. J. Math.* **9**, 1053–1060.
E GILLMAN, L. and M. JERISON (1960). *Rings of Continuous Functions*, Van Nostrand, Princeton.
F HU, S.-T. (1949). Boundedness in a topological space, *J. Math. pures appl.* **28**, 287–320.
G HURSCH, J. L. (1967). A theorem about hyperspaces, *Proc. Camb. phil. Soc.* **63**, 597–599.
H ILIADIS, S. and S. FOMIN (1966). The method of centred systems in the theory of topological spaces, *Usp. math. Nauk,* **21**, 47–76 (in Russian); English translation in *Russ. Math. Survs.* **21**, 37–62.
I ISBELL, J. R. (1966). Insufficiency of the hyperspace, *Proc. Camb. phil. Soc.* **62**, 685–686.
J KELLEY, J. L. (1961). *General Topology*, Van Nostrand, Princeton.
K MAGILL, K. D., Jr. (1965). N-point compactifications, *Am. math. Mon.* **72**, 1075–1081.
L TUKEY J. W. (1940). *Convergence and Uniformity in Topology*, Ann. Math. Stud. 2, Princeton.
M WARD, A. J. (1966). A counter-example in uniformity theory, *Proc. Camb. phil. Soc.* **62**, 207–208.
N WEIL, A. (1937). *Sur les espaces à structure uniforme et sur la topologie générale,* Actual. scient. ind. 551, Paris.

BIBLIOGRAPHY FOR PROXIMITY SPACES

(This list is intended to give all publications in Proximity Spaces and therefore not all are referred to in the text.)

1 AARTS, J. M. (1966). Dimension and deficiency in general topology, *Druk. V.R.B.* Kleine der A 3–4 Groningen.

2 ALEXANDROFF, P. (1954). Aus der Mengentheoretischen Topologie der letzten zwanzig Jahren, Proc. of the International Congress of Mathematicians, Amsterdam, I, 177–196; *MR* **20** # 2697.*

3 ALEXANDROFF, P. (1956). On two theorems of Yu. Smirnov in the theory of bicompact extensions, *Fundam. Math.* **43**, 394–398 (in Russian); *MR* **18**, 813.

4 ALFSEN, E. M. and J. E. FENSTAD (1959). On the equivalence between proximity structures and totally bounded uniform structures, *Math. Scand.* **7**, 353–360 (Correction (1961), Ibid. **9**, 258); *MR* **22** # 5958.

5 ALFSEN, E. M. and J. E. FENSTAD (1960). A note on completion and compactification, *Math. Scand.* **8**, 97–104; *MR* **23** # A 3543.

6 ALFSEN, E. M. and O. NJÅSTAD (1963). Proximity and generalized uniformity, *Fundam. Math.* **52**, 235–252; *MR* **27** # 4207a.

7 ALFSEN, E. M. and O. NJÅSTAD (1963). Totality of uniform structures with linearly ordered base, *Fundam. Math.* **52**, 253–256; *MR* **27** # 4207b.

8 BANASCHEWSKI, BERNHARD and JEAN-MARIE MARANDA (1961). Proximity functions, *Math. Nachr.* **23**, 1–37; *MR* **29** # 2768.

9 BOGNÁR, M. (1962). Bemerkungen zum Kongressvortrag 'Stetigkeitsbegriff und abstrakte Mengenlehre', von F. Riesz, *General Topology and Its Relations to Modern Analysis and Algebra*, (Proc. Sympos., Prague, 1961), Academic Press, New York, Publ. House Czech. Acad. Sci., Prague, 96–105; *MR* **27** # 2953.

10 ČECH, EDUARD (1966). *Topological Spaces*, Interscience Publishers, John Wiley and Sons.

11 CSÁSZÁR, A. (1957). Sur une classe de structures topologiques générales, *Rev. Math. pures appl.* **2**, 399–407; *MR* **20** # 1289.

12 CSÁSZÁR, A. (1963). *Foundations of General Topology*, Macmillan, New York.

13 CSÁSZÁR, A. et S. Mrówka (1959). Sur la compactification des espaces de proximité, *Fundam. Math.* **46**, 195–207; *MR* **20** # 7255.

4 DOĬČINOV, D. (1964). A method of introducing the concept of proximity. *C.R. Acad. Bulgare Sci.* **17**, 349–351 (in Russian); *MR* **29** # 4031.

* *MR* denotes Mathematical Reviews

15 DOĬČINOV, D. (1964). A unified theory of topological spaces, proximity spaces and uniform spaces, *Dokl. Akad. Nauk SSSR*, **156**, 21–24 (in Russian); *MR* **29** # 5221.

16 DOOHER, T. E. (1966). Proximity relations on an abstract lattice, Ph.D. dissertation, University of Colorado.

17 DOWKER, C. H. (1961). Mappings of proximity structures, *Proc. Symp. Gen. Top.* Prague, 139–141; *MR* **26** # 4312.

18 EFREMOVIČ, V. A. (1951). Infinitesimal spaces, *Dokl. Akad. Nauk SSSR*, **76**, 341–343 (in Russian); *MR* **12**, 744.

19 EFREMOVIČ, V. A. (1952). The geometry of proximity I, *Mat. Sb.* **31** (73), 189–200 (in Russian); *MR* **14**, 1106.

20 EFREMOVIČ, V. A. and A. S. ŠVARC (1953). A new definition of uniform spaces. Metrization of proximity spaces, *Dokl. Akad. Nauk SSSR*, **89**, 393–396 (in Russian); *MR* **15**, 815.

21 EFREMOVIČ, V. A., MORDKOVIČ, A. G. and V. J. SANDBERG, (1967). Correct spaces, *Dokl. Akad. Nauk SSSR*, **172**, 1254–1257 (in Russian); English translation in *Soviet Math. Dokl.* **8**, no. 1, 254–258; *MR* **35** # 977.

22 FEDORČUK, V. (1967). θ-space and perfect irreducible mappings of topological spaces, *Dokl. Akad. Nauk SSSR*, **174**, 757–759 (in Russian); English translation in *Soviet Math. Dokl.*, **8**, 684–686; *MR* **35** # 7288.

23 FENSTAD, J. E. (1964). On *l*-groups of uniformly continuous functions. III. Proximity spaces, *Math. Z.* **83**, 133–139; *MR* **30** # 1494.

24 FOMIN, S. V. (1958). On the connection between proximity spaces and the bicompact extension of completely regular spaces, *Dokl Akad. Nauk SSSR*, **121**, 236–238 (in Russian); *MR* **20** # 4255.

25 GACSÁLYI, SÁNDOR (1964). On proximity functions and symmetrical topogenous structures, *Publ. Math. Debrecen*, **11**, 165–174; *MR* **30** # 5271.

26 GACSÁLYI SÁNDOR (1965). Skew proximity functions, *Publ. Math. Debrecen*, **12**, 271–280; *MR* **33** # 1836.

27 GÁL, I. S. (1959). Proximity relations and precompact structures, *Indag. Math.* **21**, 304–326; *MR* **21** # 5944.

28 GOETZ, A. (1962). On a notion of uniformity for *L*-spaces of Fréchet, *Colloq. Math.* **9**, 223–231; *MR* **25** # 5491.

29 GOETZ, A. (1967). Proximity-like structures in convergence spaces, (Presented at conference on *General Topology* at Arizona State Univ.); Preprint, Univ. of Notre Dame.

30 HACQUE, M. (1962). Sur les *E*-structures, *C.R. Acad. Sci. Paris*, **254**, 1905–1907; *MR* **26** # 1848*a*.

31 HADDAD, L. (1962). Sur la notion de tramail, *C.R. Acad. Sci. Paris*, **255**, 2880–2882; *MR* **26** # 1849.

32 HAYASHI, E. (1964). On some properties of a proximity, *J. math. Soc. Japan* **16**, 375–378; *MR* **31** # 2708.

33 HAYASHI, E. (1965). A note on proximity spaces, *Bull. Aichi Gakngei Univ.* **14**, 1–4.

34 HEUCHENNE, C. (1965). Sur les connexions entre topologies et proximités, *Bull. Soc. Roy. Sci. Liège*, **34**, 215–230; *MR* **32** # 1687.

35 HEUCHENNE, C. (1967). Compléments à l'étude des familles topologiques, proximales et uniformes, *Bull. Soc. Roy. Sci. Liège*, **36**, 229–237.

36 HUNSAKER, W. N. (1967). Proximities and uniform structures induced by families of real functions, Ph.D. dissertation, Washington State University.

37 HURSCH, JACK, L., Jr. (1965). Proximity and height, *Math. Scand.* **17**, 150–160; *MR* **34** # 1982.

38 HURSCH, JACK, L., Jr. (1967). An example of two uniformities equal in height and proximity, *Proc. Am. math. Soc.* **18**, 719–722; *MR* **35** # 3627.

39 HUŠEK, M. (1964). *S*-categories, *Comment. Math. Univ. Carolinae*, **5**, 37–46; *MR* **30** # 4234.

40 HUŠEK, M. (1964). Generalized proximity and uniform spaces I, *Comment. Math. Univ. Carolinae*, **5**, 247–266; *MR* **31** # 713.

41 HUŠEK, M. (1965). Generalized proximity and uniform spaces II, *Comment. Math. Univ. Carolinae*, **6**, 119–139; *MR* **31** # 1652.

42 HUŠEK, M. (1967). Construction of special functors and its applications, *Comment. Math. Univ. Carolinae*, **8**, 555–566.

43 ISBELL, J. R. (1960). On inductive dimension of proximity spaces, *Dokl. Akad. Nauk SSSR*, **134**, 36–38 (in Russian); English translation in *Soviet Math. Dokl.* **1**, 1013–1015; *MR* **23** # A2193.

44 ISBELL, J. R. (1964). *Uniform spaces*, Mathematical Surveys, no. 12, *Am. math. Soc.*, Prov. R.I.

45 IVANOV, A. A. (1959). Contiguity relations on topological spaces, *Dokl. Akad. Nauk SSSR*, **128**, 33–36 (in Russian); *MR* **21** # 7495.

46 IVANOVA, V. M. (1964). Proximity relations and spaces of closed subsets, *Math. Sb.* **65** (107), 18–27 (in Russian); *MR* **30** # 559.

47 IVANOVA, V. M. and A. A. IVANOV (1959). Contiguity spaces and bicompact extensions of topological spaces, *Dokl. Akad. Nauk SSSR*, **127**, 20–22 (in Russian); *MR* **21** # 5943.

48 JARULKIN, N. G. (1956). Generalized infinitesimal spaces, *Ivanov. Gos. Ped. Inst. Uč. Zap. Fiz.-Mat. Nauki.* **10**, 61–79 (in Russian); *MR* **19**, 436.

49 JARULKIN, N. G. (1957). On the generalized proximity spaces, *Math. Sb.* **43** (85), 397–400 (in Russian); *MR* **20** # 2690.

50 KATĚTOV, M. (1960). Über die Berührungsräume, *Wiss. Z. Humboldt-Univ. Berlin math.-natur. R.* **9**, 685–691 (Russian, English and French summaries); *MR* **32** # 1672.

51 KATĚTOV, M. and J. VANIČEK (1964). On the proximity generated by entire functions, *Comment Math. Univ. Carolinae*, **5**, 267–278; *MR* **30** # 4235.

52 KRISHNA MURTI, S. B. (1940). A set of axioms for topological algebra, *J. Indian math. Soc.* **4**, 116–119; *MR* **2**, 69.

53 LEADER, S. (1959). On clusters in proximity spaces, *Fundam. Math.* 47, 205–213; *MR* 22 # 2978.

54 LEADER, S. (1960). On completion of proximity spaces by local clusters, *Fundam. Math.* 48, 201–216; *MR* 22 # 4047.

55 LEADER, S. (1962). On duality in proximity spaces, *Proc. Am. math. Soc.* 13, 518–523; *MR* 25 # 4486.

56 LEADER, S. (1963). On a problem of Alfsen and Fenstad, *Math. Scand.* 13, 44–46; *MR* 29 # 586.

57 LEADER, S. (1964). On products of proximity spaces, *Math. Annalen*, 154, 185–194; *MR* 28 # 5420.

58 LEADER, S. (1965). On pseudometrics for generalized uniform structures, *Proc. Am. math. Soc.* 16, 493–495; *MR* 31 # 714.

59 LEADER, S. (1967). Local proximity spaces, *Math. Annalen*, 169, 275–281.

60 LEADER, S. (1967). Spectral structures and uniform continuity, *Fundam. Math.* 60, 105–115; *MR* 35 # 974.

61 LEADER, S. (1967). Metrization of proximity spaces, *Proc. Am. math. Soc.* 18, 1084–1088.

62 LEADER, S. (1968). On metrizable precompact proximity spaces, *Nieuw Arch. Wisk.* Series 3, 16, 12–14.

63 LODATO, M. W. (1964). On topologically induced generalized proximity relations, *Proc. Am. math. Soc.* 15, 417–422; *MR* 28 # 4513.

64 LODATO, M. W. (1966). On topologically induced generalized proximity relations II, *Pacif. J. Math.* 17, 131–135; *MR* 33 # 695.

65 LODATO, M. W. (1966). Generalized proximity spaces: A generalization of topological spaces, (unpublished).

66 MAMUZIČ, Z. P. (1963). *Introduction to General Topology*, P. Noordhoff Ltd. Groningen, The Netherlands.

67 MATTSON, D. A. (1967). Separation relations and quasi-proximities, *Math. Annalen*, 171, 87–92; *MR* 35 # 3628.

68 MEENAKSHI, K. N. (1966). Proximity structures in Boolean algebras, *Acta Sci. Math. (Szeged)*, 27, 85–92; *MR* 33 # 4893.

69 MORDKOVIČ, A. G. (1965). Systems with small sets and proximity spaces. *Mat. Sb.* 67 (109), 474–480 (in Russian); *MR* 32 # 8310.

70 MORDKOVIČ, A. G. (1966). A criterion for the correctness of a uniform space, *Dokl. Akad. Nauk SSSR*, 169, 276–279 (in Russian); English translation in *Soviet Math. Dokl.* 7, 915–918; *MR* 33 # 4894.

71 MOREZ, N. S. (1965). Generators for the semi-topogenous orders of Császár, Ph.D. dissertation, University of Colorado.

72 MOZZOCHI, C. J. (1968). Symmetric generalized uniform and proximity spaces, Ph.D. dissertation, University of Connecticut.

73 MRÓWKA, S. G. (1956). On the notion of completeness in proximity spaces, *Bull. Acad. Polon. Sci. Cl. III*, 4, 477–478; *MR* 19, 668.

74 MRÓWKA, S. G. (1956). On complete proximity spaces, *Dokl. Akad. Nauk SSSR*, 108, 587–590 (in Russian); *MR* 19, 158.

75 Mrówka, S. G. (1957). On the uniform convergences in proximity spaces, *Bull. Acad. Polon. Sci. Cl.* III, **5**, 255–257; *MR* **19**, 669.
76 Mrówka, S. G. (1960). Axiomatic characterization of the family of all clusters in a proximity space, *Fundam. Math.* **48**, 123–126; *MR* **22** # 7107.
77 Mrówka, S. G. and W. J. Pervin (1964). On uniform connectedness, *Proc. Am. math. Soc.* **15**, 446–449; *MR* **28** # 4515.
78 Myškis, A. D. (1952). On the relation of infinitesimal spaces with extensions of topological spaces, *Dokl. Akad. Nauk SSSR*, **84**, 879–882 (in Russian); *MR* **14**, 1001.
79 Myškis, A. D. and E. I. Vigant (1955). On a connection of proximity spaces with extensions of topological spaces, *Dokl. Akad. Nauk SSSR*, **103**, 969–972 (in Russian); *MR* **18**, 140.
80 Njåstad, O. (1963). Some properties of proximity and generalized uniformity, *Math. Scand.* **12**, 47–56; *MR* **28** # 2523.
81 Njåstad, O. (1964). On real-valued proximity mappings, *Math. Annalen*, **154**, 413–419; *MR* **29** # 4030.
82 Njåstad, O. (1965). On uniform spaces where all uniformly continuous functions are bounded, *Monatsh. Math.* **69**, 167–176; *MR* **31** # 2710.
83 Njåstad, O. (1966). On Wallman-type compactifications, *Math. Z.* **91**, 267–276; *MR* **32** # 6404.
84 Pervin, W. J. (1963). Quasi-proximities for topological spaces, *Math. Annalen*, **150**, 325–326; *MR* **27** # 1917.
85 Pervin, W. J. (1964). On separation and proximity spaces, *Am. math. Mon.* **71**, 158–161; *MR* **29** # 2767.
86 Pervin, W. J. (1964). Equinormal proximity spaces. *Indag. Math.* **26**, no. 2, 152–154; *MR* **28** # 4514.
87 Poljakov, V. Z. (1964). Regularity, product and spectra of proximity spaces, *Dokl. Akad. Nauk SSSR*, **154**, 51–54 (in Russian); English translation in *Soviet Math. Dokl.* **5**, 45–49; *MR* **28** # 582.
88 Poljakov, V. Z. (1964). Open mappings of proximity spaces, *Dokl. Akad. Nauk SSSR*, **155**, 1014–1017 (in Russian); *MR* **30** # 2460.
89 Poljakov, V. Z. (1965). Regularity and the product of proximity spaces, *Mat. Sb.* **67** (109), 428–439 (in Russian); *MR* **32** # 8311.
90 Poljakov, V. Z. (1965). On the regularity of the proximity product of regular spaces, *Mat. Sb.* **68** (110), 242–250 (in Russian); *MR* **32** # 6405.
91 Poljakov, V. Z. (1965). On uniform convergence spaces, *Colloquium Math.* **13**, 167–179 (in Russian).
92 Ramm, N. S. and A. S. Švarc (1953). Geometry of proximity, uniform geometry and topology, *Mat. Sb.* N.S. **33** (75), 157–180, (in Russian); *MR* **15**, 815.
93 Reed, E. E. (1968). Uniformities obtained from filter spaces, (unpublished).
94 Reed, E. E. and W. J. Thron (1969). *m*-bounded uniformities between two given uniformities, *Trans. Am. math. Soc.* **141**, 71–77; *MR* **39** # 4805.

122   BIBLIOGRAPHY

95 RIESZ, F. (1908). Stetigkeitsbegriff und abstrakte Mengenlehre, *Atti IV Congr. Intern. Mat. Roma*, II, 18–24.

96 SIEBER, J. L. and W. J. PERVIN (1964). Connectedness in syntopogenous spaces, *Proc. Am. math. Soc.* **15**, 590–595; *MR* **29** # 4020.

97 SMIRNOV, Y. M. (1952). On proximity spaces in the sense of V. A. Efremovič, *Dokl. Akad. Nauk SSSR*, **84**, 895–898 (in Russian); English translation in *Am. math. Soc. Transl.* Ser. 2, **38**, 1–4; *MR* **14**, 1107.

98 SMIRNOV, Y. M. (1952). On proximity spaces, *Mat. Sb.* **31** (73), 543–574 (in Russian); English translation in *Am. math. Soc. Transl.* Ser. 2, **38**, 5–35; *MR* **14**, 1107.

99 SMIRNOV, Y. M. (1953). On the completeness of proximity spaces, *Dokl. Akad. Nauk SSSR*, **88**, 761–764 (in Russian); *MR* **15**, 144.

100 SMIRNOV, Y. M. (1953). On completeness of uniform spaces and proximity spaces. *Dokl. Akad. Nauk SSSR*, **91**, 1281–1284 (in Russian); *MR* **16**, 58.

101 SMIRNOV, Y. M. (1954). On the dimension of proximity spaces, *Dokl. Akad. Nauk SSSR*, **95**, 717–720 (in Russian); *MR* **16**, 845.

102 SMIRNOV, Y. M. (1956). On the dimension of proximity spaces, *Mat. Sb.* **38** (80), 283–302 (in Russian); English translation in *Am. math. Soc. Transl.* Ser. 2, **21**, 1–20; *MR* **27** # 715.

103 SMIRNOV, Y. M. (1956). Geometry of infinite uniform complexes and δ-dimensionality of point sets, *Mat. Sb.* **40** (82), 137–156 (in Russian); English translation in *Am. math. Soc. Transl.* Ser. 2, **15**, 95–113; *MR* **22** # 5960.

104 SMIRNOV, Y. M. (1962). Some questions of uniform topology, *Proc. Internat. Congr. Mathematicians, Stockholm*, 497–501; *MR* **31** # 715.

105 SMIRNOV, Y. M. (1964). On the completeness of proximity spaces I, *Trudy Moskov Mat. Obšč*, **3**, 271–306 (in Russian); English translation in *Am. math. Soc. Transl.* Ser. 2, **38**, 37–73; *MR* **16**, 844.

106 SMIRNOV, Y. M. (1964). On the completeness of proximity spaces II, *Trudy Moskov Mat. Obšč*, **4**, 421–438 (in Russian); English translation in *Am. math. Soc. Transl.* Ser. 2, **38**, 75–93; *MR* **17**, 286.

107 SMIRNOV, Y. M. (1966). On the dimension of remainders in bicompact extensions of proximity and topological spaces, *Mat. Sb.* **69** (111), 141–160 (in Russian); *MR* **33** # 6579.

108 SMIRNOV, Y. M. (1966). Dimension of increments of proximity spaces and of topological spaces, *Dokl. Akad. Nauk SSSR*, **168**, 528–531; *MR* **35** # 4886.

109 SMITH, D. HAMMOND (1966). Hyperspaces of a uniformizable space, *Proc. Camb. phil. Soc.* **62**, 25–28; *MR* **32** # 4654.

110 STEINER, E. F. (1964). The relation between quasi-proximities and topological spaces, *Math. Annalen*, **155**, 194–195; *MR* **29** # 581.

111 STEVENSON, F. W. (1966). Uniform spaces with linearly ordered bases, Ph.D. dissertation, University of Colorado.

112 ŠVARC, A. S. (1956). Proximity spaces and lattices, *Učen. Zap. Ivanovsk Gos. Ped. Inst.* **10**, 55–60 (in Russian); *MR* **19**, 436.

113 SZYMANSKI, P. (1941). La notion des ensembles séparés comme terme primitif de la topologie, *Math. Timisoara*, **17**, 65–84; *MR* **4**, 87.

114 TERWILLIGER, W. L. (1965). On contiguity spaces, Ph.D. dissertation, Washington State University.

115 THRON, W. J. (1966). *Topological Structures*, Holt, Rinehart and Winston, New York.

116 WALLACE, A. D. (1941). Separation spaces, *Annals of Math.* **42**, 687–697; *MR* **3**, 57.

117 WALLACE, A. D. (1942). Separation spaces II, *Anais. Acad. Brasil Ciencias*, **14**, 203–206; *MR* **4**, 87.

118 WARD, A. J. (1967). On *H*-equivalence of uniformities; the Isbell–Smith problem, *Pacif. J. Math.* **22**, 189–196.

119 WARD, A. J. (1969). On *H*-equivalence of uniformities II, *Pacif. J. Math.* **28**, 207–215.

120 WARRACK, B. D. and S. A. NAIMPALLY (1968). Clusters and ultrafilters, *Pub. L'Inst. Math.* **8** (22), 100–101.

121 WARRACK, B. D. and S. A. NAIMPALLY (1968). On completion of a uniform space by Cauchy clusters, Contributions to *Extension Theory of Topological Structures* (Proc. Sympos. Berlin 1967), VEB Deutscher Verlag Der Wissenschaften, Berlin.

122 WOLK, E. S. (1968). Convergence in proximity in partially ordered sets, (unpublished).

123 Zaicev, V. I. (1967). On the theory of Tychonoff spaces, *Vestnik Moskov. Univ. Ser. I Mat. Meh.* **22**, 48–57; *MR* **35** # 7286.

Supplement

124 DOĬČINOV, D. (1969). A generalization of topological spaces, Contributions to *Extension Theory of Topological Structures* (Proc. Sympos. Berlin 1967), VEB Deutscher Verlag Der Wissenschaften, Berlin.

125 ENGELKING, R. (1968). *Outline of General Topology*, North-Holland Publishing Company, Amsterdam.

126 FEDORČUK, V. (1968). *θ*-proximities and *θ*-absolutes, *Soviet Math. Dokl.* **9**, 661–664.

127 HADZIIVANOV, N. (1969). Pseudometrics of proximity spaces, Contributions to *Extension Theory of Topological Structures* (Proc. Sympos. Berlin 1967), VEB Deutscher Verlag Der Wissenschaften, Berlin.

128 HAYASHI, E. (1967). Closures and neighbourhoods in certain proximity spaces, *Proc. Japan Acad.* **43**, 619–623.

129 HUŠEK, M. (1969). Categorical connections between generalized proximity spaces and compactifications, Contributions to *Extension Theory of Topological Structures* (Proc. Sympos. Berlin 1967), VEB Deutscher Verlag Der Wissenschaften, Berlin.

130   LEADER, S. (1969). Extensions based on proximity and bounded-
      ness, *Math. Z.* **108**, 137–144.
131   MOZZOCHI, C. J. (1968). A partial generalization of a theorem of
      Hursch, *Comment. Math. Univ. Carolinae*, **9**, 157–159.
132   MOZZOCHI, C. J. (1969). *Symmetric generalized uniform and proxi-
      mity spaces*, Trinity College, Hartford, Connecticut.
133   NACHMAN, L. J. (1968). Weak and strong constructions in proxi-
      mity spaces, Ph.D. dissertation, The Ohio State University.
134   NAGATA, J. (1968). *Modern General Topology*, North-Holland.
135   POLJAKOV, V. Z. (1969). On some proximity properties determined
      only by the topology of the compactification, Contributions to
      Extension Theory of Topological Structures (Proc. Sympos.
      Berlin 1967), VEB Deutscher Verlag Der Wissenschaften, Ber-
      lin.
136   SMIRNOV, Y. M. (1952). Mappings of systems of open sets, *Mat.
      Sb.* **31** (73), 152–166 (in Russian); *MR* **14**, 303.
137   SMIRNOV, Y. M. (1960). Generalization of the Stone–Weierstrass
      theorem and proximity spaces, *Czech. Math. J.* **10** (85), 493–500
      (in Russian, English summary); *MR* **23** # A 3474.
138   GAGRAT, M. S. and S. A. NAIMPALLY (1970). Proximity approach
      to extension problems, *Fundam Math.* (to appear).

INDEX

uniformity (*cont.*)
  total, A-N, 79
  universal, 71
uniformity base, 64
uniformity subbase, 64
uniformly continuous, 22, 66
uniformly finer, 89

universal uniformity, 71
Urysohn's lemma, analogue of, 44

weight
  proximity, 48
  topological, 47